COMPLEXITY IN CHEMISTRY

Mathematical Chemistry

A series of books edited by:

Danail Bonchev, Department of Marine Sciences, Texas A&M University, Galveston, Texas, USA
Dennis H. Rouvray, Department of Chemistry, University of Georgia, Athens, Georgia, USA

COMPLEXITY IN CHEMISTRY
Introduction and Fundamentals

Edited by

Danail Bonchev
Department of Marine Sciences, A&M University
Galveston, Texas, USA

and

Dennis H. Rouvray
Department of Chemistry, University of Georgia,
Athens, Georgia, USA

Taylor & Francis
Taylor & Francis Group
LONDON AND NEW YORK

First published 2003
by Taylor & Francis
11 New Fetter Lane, London EC4P 4EE

Simultaneously published in the USA and Canada
by Taylor & Francis Inc.
29 West 35th Street, New York, NY 10001

Taylor & Francis is an imprint of the Taylor & Francis Group

Typeset in Times by
HWA Text and Data Management, Tunbridge Wells
Printed and bound in Great Britain by
The Cromwell Press, Trowbridge, Wiltshire

British Library Cataloguing in Publication Data
A catalogue record for this book is available from the British Library

Library of Congress Cataloging in Publication Data
A catalog record for this book has been requested

ISBN 0–415–28791–X

CONTENTS

INTRODUCTION TO THE SERIES

The mathematization of chemistry has a long and colorful history extending back well over two centuries. At any period in the development of chemistry the extent of the mathematization process roughly parallels the progress of chemistry as a whole. Thus, in 1786 the German philosopher Immanuel Kant observed [1] that the chemistry of his day could not qualify as one of the natural sciences because of its insufficient degree of mathematization. It was not until almost a century later that the process really began to take hold. In 1874 one of the great pioneers of chemical structure theory, Alexander Crum Brown (1838–1922), prophesied [2] that "... chemistry will become a branch of applied mathematics; but it will not cease to be an experimental science. Mathematics may enable us retrospectively to justify results obtained by experiment, may point out useful lines of research and even sometimes predict entirely novel discoveries. We do not know when the change will take place, or whether it will be gradual or sudden ..." This prophecy was soon to be fulfilled. Indeed, even before these words were uttered, combinatorial methods were being employed for the enumeration of isomeric species [3]. During Crum Brown's lifetime algebraic equations were used to predict the properties of substances, calculus was employed in the description of thermodynamic and kinetic behavior of chemical systems, and graph theory was adapted for the structural characterization of molecular species.

During the twentieth century the applications of mathematics came thick and fast. The advent of quantum chemistry in the 1920s brought in its wake a host of mathematical disciplines that chemists felt obliged to master. These included several areas of linear algebra, such as matrix theory and group theory, as well as calculus. Group theory has become so widely accepted by chemists that it is now used routinely in areas such as crystallography and molecular structure analysis. Graph theory seems to be following in the footsteps of group theory and is currently being exploited in a wide range of applications involving the classification, systemization, enumeration and design of systems of chemical interest. Topology has found important applications in areas as diverse as the characterization of potential energy surfaces, the discussion of chirality, and the description of catenated and knotted molecular species. Information theory has yielded valuable insights into the nature of thermodynamic processes and the origin of life. The contemporary fascination with dissipative systems, fractal phenomena and chaotic behavior has again introduced new mathematics, such as catastrophe theory and fractal geometry, to the chemist.

All of these and numerous other applications of mathematics that have been made in the chemical domain have brought us to a point where we consider it may now be fairly said that mathematics plays an indispensable role in modern chemistry. Because of the burgeoning use of mathematics by chemists and the current feeling that mathematics is opening up some very exciting new directions to explore, we believe that the 1990s represented a particularly auspicious time to present a comprehensive treatment of the manifold applications of mathematics to chemistry. We were persuaded to undertake this somewhat awesome task after much reflection and eventually decided to publish our material in a series of volumes, each of which is to be devoted to a discussion of the applications of a specific branch of mathematics. The title of our series, *Mathematical Chemistry*, was chosen to reflect as accurately as possible the proposed contents. The term 'mathematical chemistry' was coined in the early 1980s to designate the field that concerns itself with the novel and nontrivial application of mathematics to chemistry. Following the usual practice in this area, we shall interpret chemistry very broadly to include not only the traditional disciplines of inorganic, organic and physical chemistry but also their hybrid offspring such as chemical physics and biochemistry.

It is anticipated that each of the volumes in our series will contain five to six separate chapters, each of which will be authored by a leading expert in the respective field. Whenever it is evident that one such volume is insufficient to do justice to a wealth of subject matter, additional volumes devoted to applications of the same branch of mathematics will be published. In this way it is hoped that our coverage will indeed be comprehensive and reflect significant developments made up to the end of the twentieth century. Our aim will be not only to provide a background survey of the various areas we cover but also to discuss important current issues and problems, and perhaps point to some of the major trends that might be reasonably expected in mathematical chemistry in the first half of the new millennium. In the first few volumes of our series we propose to examine the applications to chemistry of graph theory, group theory, topology, combinatorics, information theory and artificial intelligence.

It may be of interest to observe here that mathematical chemists have often applied and even sought after branches of mathematics that have tended to be overlooked by the chemical community at large. This is not to imply that the mathematics itself is necessarily new—in fact, it may be quite old. What is new is the application to chemistry; this is why the word novel was employed in our earlier definition of mathematical chemistry. The thrill of discovering and developing some novel application in this sense has been an important source of motivation for many mathematical chemists. The other adjective used in our definition of mathematical chemistry, i.e., non-trivial, is also worthy of brief comment. To yield profitable new insights, the mathematics exploited in a chemical context usually needs to be of at least a reasonably high level. In an endeavor to maintain a uniformly high level, we shall seek to ensure that all of the contributions to our volumes are written by researchers at the forefront of their respective disciplines. As a consequence, the contents of our various volumes are likely to appeal to a fairly sophisticated audience: bright undergraduate and post-graduate students, researchers operating at the tertiary level in academia, industry or government service, and perhaps even to newcomers to the area desirous of experiencing an invigorating excursion through the realms of mathematical chemistry. Overall, we hope that our series will provide a valuable resource for scientists and mathematicians seeking an authoritative and detailed account of mathematical techniques to chemistry.

In conclusion, we would like to take this opportunity of thanking all our authors, both those who have contributed chapters so far and those who have agreed to submit contributions for forthcoming volumes. It is our sincere hope that the material to be presented in our series will find resonance with our readership and afford many hours of enjoyable and stimulating reading.

Danail Bonchev
Dennis H. Rouvray

1. I. Kant, *Metaphysische Anfangsgründe der Naturwissenschaft,* Hartknoch Verlag, Riga, 1786.
2. A. Crum Brown, *Rept. Brit. Assoc. Adv. Sci.* 45–50 (1874).
3. F.M. Flavitsky, *J. Russ. Chem. Soc.* **3**, 160 (1871).

PREFACE

The present volume constitutes Volume 7 of our ongoing series of monographs on various themes in the broad field of mathematical chemistry. To date, Volumes 1 and 2 have considered graph theory and its applications in the chemical context, Volumes 3 and 4 treated group theory and its chemical applications, and Volumes 5 and 6 have looked at the chemical applications of topology. In this volume the topic we have selected for coverage is the rapidly developing and now burgeoning field of complexity studies. This field in a very real sense unites all of the scientific disciplines and even establishes important links with fields usually described as the arts and the humanities. The reason is not hard to find: our world and indeed the entire universe in which we live may be interpreted as being complex. In fact, it is very difficult, if not impossible, to find anything in the natural world that could be characterized as simple if it is examined in sufficient detail. Certainly atoms, which were once thought to be tiny chunks of matter, are now realized to be highly complex worlds in their own right and even the subatomic particles are currently considered by many to be possessed of complex structures.

In our opening Chapter 1 the scene is set for what follows in the rest of the book. The underlying theme is that complexity is to be found everywhere and that, once one is trained to look for it, complexity appears in all manner of guises. Because it is so ubiquitous, the terms complex and complexity have come to be associated with many different meanings and this fact makes it quite difficult to pin down what we are to understand when these terms are used. There would appear to be general agreement that, although these terms can be defined precisely in specific instances, the terms are understood intuitively. Like most other chemical concepts, therefore they are essentially fuzzy in nature. This fact, however, in no way diminishes their great usefulness in chemistry. Chemical applications abound and a number of the more important ones are touched upon in this chapter. Thus, the role of thermodynamic entropy in the formation of highly complex structures is explored. It is pointed out that such structures can be characterized as reservoirs of negative entropy and that they can exist only in systems that are far from equilibrium. Thus, living systems, which are prime examples of complex, self-organized systems, exist at the edge of chaos. It is this circumstance that renders possible all of the almost incredible range of structures and diversity encountered in the natural world. Without the existence of self-organizing complex systems none of us would be here to tell the tale.

The themes broached in the opening chapter are explored in much greater depth in succeeding chapters in our monograph. Thus, Chapter 2, which is written by a group of leading experts from Croatia and Germany, addresses the theme of molecular complexity indices. These authors offer a comprehensive analysis of some of the most significant work carried out in complexity studies within a strictly chemical context.

A critical comparison is made of the criteria to be adopted for the establishment of appropriate complexity measures as applied to molecular species. The measures put forward over the past quarter of a century are described and assessed. Not only is recent work on molecular complexity indices discussed but this is set into context by tracing the early development of such indices. Mention is made of the Minoli combinatorial complexity, the Bonchev–Trinajstić distance information index, Bertz's molecular complexity, the number of spanning trees, the number of connected subgraphs, Bonchev's overall connectivity index, the Zagreb complexity indices, Randić's augmented connectivity, and the molecular labirinthicity of Rücker and Rücker. Comparisons are made between the complexity indices reviewed for selected classes of acyclic and cyclic molecular species. The ability of these descriptors to capture the major patterns of molecular branching and cyclicity are explored. An essential difference between the behavior of symmetry-based and symmetry-independent complexity measures is highlighted.

In Chapter 3 Bertz involves the reader in a vivid discussion of what is meant by complexity and the various ways in which it might be measured. This author has been particularly concerned with the development of complexity indices that are relevant for the characterization of chemical synthesis mechanisms. Here he presents original views on the classification of complexity into intrinsic and extrinsic forms as well as constructive and instructive kinds of complexity. The role of a number of relevant complexity factors is assessed and their formation into a hierarchy is explored. Not surprisingly, special attention is devoted to complexity measures developed by the author himself and these include the total number of substructures, the number of types of substructures, and also a series of indices based on graph-theoretical cliques and bicliques. After his overview of such complexity measures, the author goes on to present details of his own extensive work on the application of complexity measures in the area of organic synthesis. He also offers a highly informative discussion on the concept of synthetic complexity.

Chapter 4 by Bonchev is devoted to a detailed treatment of complexity measures that are based on Shannon's information theory. This theory is viewed in the present context as an extension of Boltzmann's statistical thermodynamics to the various processes of transmitting information. The intimate connection between information and entropy is also addressed. The evolution of the concept of the information content of a molecule is traced from the earliest pioneering works of the 1950s to the most recent studies. A critical analysis is put forward on the usage of the symmetry-based Shannon function as a measure of complexity. The advantage of the magnitude-based information-theoretic indices is pointed out by proceeding from a concept elaborated by Bonchev and Trinajstić in 1977. A number of newer developments of this concept involve its combination with complexity measures based on all the substructures in a species, including Bonchev's overall complexity measures, i.e. overall connectivity, the overall Wiener index, and the overall Zagreb indices.

The final Chapter 5 by Seitz on the subject of thermodynamic complexity continues with the thermodynamic theme and the interconnections between information and thermodynamics advanced in the previous chapter. After presenting a review of work, the author discusses the division of complexity studies into two weakly interacting areas, namely the complexity of objects and the complexity of processes. The two areas are analyzed within the chemical context and the process complexity is treated from an original thermodynamic point of view. Maxwell's Demon is reincarnated to permit the introduction of a new thermodynamic state function appropriate for the study of complexity. This function is related to the microscopic (or internal) work performed in the system. The thermodynamic complexity is also defined for near-equilibrium steady state ideal gases subjected to thermal gradients.

We would like to take this opportunity of thanking all of our contributors to this volume for their highly stimulating contributions to the fascinating topic of complexity applied in the chemical context. It has been an enjoyable experience for us to work and collaborate with them and we can only hope that our readers will derive as much pleasure as we have when they begin to peruse the exciting offerings contained herein.

Danail Bonchev
Dennis H. Rouvray

Chapter 1

AN INTRODUCTION TO COMPLEXITY

D.H. Rouvray
Department of Chemistry, University of Georgia,
Athens, Georgia 30602-2556, USA

Everything that is simple is merely imaginary, is not "true." What is real and what is true, however, is neither simple nor is it reducible to simplicity.

Friedrich Nietzsche

1.1 The Opening Scenario

Because the terms *complex* and *complexity* are so widely used in everyday language, when an object is described as being complex or a system is said to exhibit complexity, most of us have an intuitive feeling that we understand what is meant. Yet, if put to the test and called upon to give a rigorous elucidation of what we believe we understand, we soon realize how difficult it is to furnish a meaningful account. In such a situation, we find ourselves in the same kind of dilemma experienced by St. Augustine when he confronted the problem of defining time. In the end he could do no better than to admit [1] that "if nobody asks me, I know; but if I were desirous to explain it to one that should ask me, plainly I know not." Both time and complexity are rather fuzzy, abstract concepts and our comprehension of them is largely independent of reason. This implies that our grasp of their meaning is founded on shifting sands because any intuitive understanding of the world around us can change quite profoundly as our insights into natural phenomena become ever more penetrating. In the words of the quantum theoretician Holland [2], intuition "is a function of history and not eternally frozen. The notion that a body persists in a state of uniform motion unless acted upon by a resultant force would be counter-intuitive to Aristotle but natural for Galileo. [Similarly,] quantum phenomena require the creation of quantum intuition."

The concept of complexity is thus an elusive and enigmatic one that presents us with a number of problems. In addition to the disconcerting problem that our intuitions about the physical world are continually being updated in the light of new interpretations of reality, we encounter several other difficulties in attempting to arrive at a workable definition of complexity. Any concept that has been applied in a wide variety of contexts necessarily becomes broadened in its scope and less well delineated in its meaning. Over and above this, the past two decades have witnessed a growing trend away from the idea that concepts in general are possessed of well-defined features and structures. The current perception is that most concepts are ill-defined and fuzzy, the only exceptions being logical or mathematical concepts that are defined in purely abstract terms [3]. It has been claimed [4] that the only characteristic that renders a concept into a recognizable unity is the rather vague family resemblance that exists among the various members included within the concept. Against such a backdrop, it is hardly surprising that differing complexity researchers have arrived at divergent interpretations of their domain. Because of the many different approaches to complexity that have been adopted, we shall need to explore a variety of avenues in an endeavor to arrive at a reasonably comprehensive understanding of the present status of complexity studies.

Before delving into any of the more technical meanings that have been assigned to the term complexity, we commence our discussion by considering how the term has traditionally been defined. To do this we turn to definitions available in well-known dictionaries. For instance, the *Comprehensive Dictionary* of Webster informs us that complexity is the state of being complex, a definition that helps us little until we learn that complex means "consisting of various parts or elements." The recent (1998) *Oxford English Dictionary* takes us somewhat further in that it points out that complex means "consisting of many different and connected parts." Thus, we may note that, in addition to being comprised of multiple parts, a complex system is deemed to have its parts interacting or, at the very least, connected to one another. With this basic idea as our starting point, we can begin to probe some of the ways in which the term has been employed in a more technical sense. Before doing this, however, we propose to examine an assumption that is all too commonly made when complex systems are considered. In virtually all cases, complex systems are assumed to be classical systems, by which is meant systems that can be studied in isolation from the external world. Such an assumption is clearly invalid because it takes no account of certain fundamental quantum-mechanical principles that disallow the idea that any system can be regarded as completely isolated from the rest of the universe.

1.2 The Holistic Backdrop

Among the philosophers of ancient Greece were some, notably Parmenides and Plotinus, who argued that our world could not be as simple as it might seem. Their contention was that the world could never be totally separated into noninteracting parts since any conceivable separation process would always leave the parts connected together in some way. Thus, they maintained that the world is intrinsically holistic in nature and that to reflect reality accurately a holistic frame of reference had to be adopted [5]. Modern quantum physics espouses a very similar view, and some of the remarkable inferences that can be drawn when quantum ideas are introduced will now be explored. We know that it is not possible to make a so-called Heisenberg cut into any part of the fabric of our universe, i.e., to effect an excision that leaves the part completely disconnected from its surroundings [6]. This in turn implies that any system we choose to study will automatically be more complex than it may seem because of the existence of invisible, yet inseverable links between every part of the system and the rest of the universe. The nature of these links and a mathematical characterization of them was first set out in a celebrated paper published in 1935 by Einstein, Podolsky and Rosen [7]. The links arise because of the omnipresence of the superposition principle which postulates that any set of wave functions used to represent the state of a system can always be superimposed upon one another to yield a new state of that system.

Because of the unbreakable links that obtain between all the parts of a system and the rest of the universe, it is evident that complete disentanglement of any system from its surroundings is not feasible. Furthermore, if we focus on the system itself, this too still needs to be characterized in holistic terms. Indeed, it has been remarked [8] that the constituents of a system are controlled by "something that cannot be described solely in terms of … parts and preassigned relationships." It is now widely accepted that the state of a system is in fact determined by the many-body wave function employed to represent it. This wave function evolves in time in a manner specified by the Schrödinger equation [9]. It is implicit within this equation that the individual parts of a system will retain their interconnectedness even when separated from one another by a great distance. In the circumstances, the best that can be accomplished when attempting to characterize a system is to effect an approximate separation of the system into its component parts. This can be achieved by performing a factorization of the total wave function, an operation that is thought to be justifiable in many of the situations extant at the current stage of development of the universe [10]. For certain purposes it is thus acceptable to carry out analyses of systems within the traditional reductionist framework of science. We also might mention that near to the classical limit the nonlocal interactions we have been discussing fade into comparative insignificance.

At this point we shall take a brief look at the largest system of all, namely the universe itself. The currently most favored theory on the origins of the universe hypothesizes that the universe was formed some 15 billion years ago. The event that produced the universe is commonly referred to as the Big Bang and this event was supposedly triggered by a random quantum fluctuation [11]. Since the time of its formation, the universe is believed to have undergone continual change and has evolved from a state of featureless simplicity through a succession of increasingly complex states to its present state. Although the reason for this evolution has remained unclear, it has been conjectured that it could be due to an innate ability of the universe to self-organize itself. In the words of Davies [12], for instance, it is argued that "nature has creative power and is able to produce a progressively richer variety of complex forms and structures." The philosopher Popper [13] has ventured even further in this direction by asserting that "[t]he greatest riddle of cosmology may well be … that the universe is, in a sense, creative." That our universe continues to change and evolve cannot be doubted; it is evidenced by the fact that all of the innumerable different objects and entities currently in existence in our world were formed since the Big Bang. According to Davies [14], our world has risen "out of featurelessness, growing rather than dying, developing new structures, processes and potentialities all the time, unfolding like a flower."

1.3 The Self-Assembly Process

The unfolding or developmental processes referred to have the effect of increasing the richness and variety of the complex forms encountered in our world. Such processes are spontaneous and are described as either self-assembly or self-organization processes. Processes of this type are manifested in numerous different situations and guises. In its beginnings, the universe is conjectured to have been extremely simple [15], for at the time of the Big Bang is it thought that there were no particles in existence (since energy and mass were interchangeable), the four basic forces of nature were still united, and no measurable events were occurring. Yet, as soon as the galaxies started to form, major increases in the complexity of the emerging universe became apparent. A typical galaxy contains around 10^{11} stars and has a diameter of 10^{23} cm and a mass of 10^{45} g. Moreover, galaxies are not distributed uniformly throughout space but occur rather as irregular clusters, sheets and strings. In addition to this, galactic formations are frequently interconnected or tangled together in very complicated ways. Perhaps not surprisingly, it has been commented [16] that when "[l]istening to astronomers talking about emerging populations, evolution and variations that become extinct, it is sometimes hard to remember that they are indeed astronomers talking about galaxies, not biologists talking about the evolution of life on Earth."

Turning now to actual living systems, we would point out that the process of evolution would not be possible if the only kind of interactions taking place in the world were random atomic and molecular collisions. The evolution of life has always occurred in one specific direction, namely from the less complex to the more complex. The formation over time of an enormous multiplicity of exceedingly complex living systems can be accounted for only in terms of self-organization processes. Let us consider why this must be so, following arguments set forth by Kauffman [17]. We start by conjecturing that the time available for the emergence of living systems on Earth is roughly four billion years, i.e., the age of the Earth.

Such a period of time would allow for an estimated 10^{51} separate attempts to be made to produce life on Earth. Suppose now that we wish to construct a single bacterium from the set of 20 amino acids that are used to generate enzymes. A typical bacterium might contain some 2000 functioning enzymes, each of which is comprised of 200 amino acids. The probability of constructing the first enzyme correctly from 20 different amino acids will be 1 in 20^{200}. However, it has been argued that this probability might be increased to around 1 in 10^{20} because of the possibility that more than one amino acid sequence could catalyze a given reaction. To assemble together correctly all of the 2000 enzymes in the bacterium is thus associated with a probability of 1 in $10^{20\times2000}$ or 1 in $10^{40,000}$. Bearing in mind that the number of separate attempts to produce living systems is only 10^{51}, it immediately becomes apparent that the construction of a single bacterium in the manner described is an unimaginably improbable event. Our conclusion must be that life could not have been formed in this way.

The pinnacle of self-organized structures, and the most complex and sophisticated machine in the world, is the human brain. In spite of its being an organ of comparatively low mass—the average adult male brain weighs around 1.40 kg and that of the adult female 1.25 kg—the brain is able to perform a host of astonishing feats [18]. One example is its capability of achieving consciousness, i.e., awareness of itself and its environment. It now seems possible to explain consciousness in terms of the underlying structure of the brain. The basic unit of the brain is the nerve cell or neuron and the human brain has between 1.0×10^{10} and 1.5×10^{10} such cells. The neurons are linked together by structures known as synapses to form prodigiously complex neural networks. Although the number of synapses approaches 10^{14}, this pales in comparison to the number of different ways in which the neurons can be interconnected—a number so large that it exceeds the number of atoms in the universe [18]. Consciousness arises within these neural networks as a highly distributed process characterized by an ever-changing sequence of states. These states are both wholly integrated and greatly differentiated [19]. Integration occurs in the sense that each state is experienced as a whole and cannot be broken down into its constituent parts. Differentiation implies that numerous different states can be experienced within a fraction of a second. The brain selects states with extreme rapidity by the exercise of imagination and by searching its memory database built up from previous experiences.

1.4 The Thermodynamic Entropy

Self-organization processes encompass among their most phenomenal feats the generation, maintenance, and evolution of living systems. Given this reality, we are prompted to ask whether the creative force behind self-organizing processes in any way conflicts with the second law of thermodynamics. This law has been formulated in more than twenty different ways [20], two of which state that:

1. All physical processes are irreversible because some of the energy in the system will always be dissipated in the form of heat.
2. In any isolated system, the entropy of that system will always increase during an irreversible process.

The terminology used here calls for some elucidation as it derives from a highly idealized model of reality. The second law refers only to systems isolated from the rest of the universe, namely to systems that do not exchange matter, energy or information with their surroundings. As we have pointed out in Section 1.2, such a system represents no more than a fictional idealization of the real world because there is always some interaction, albeit very small, between a system and its environment. In addition to the inseverable holistic links we discussed earlier, it would appear to be virtually impossible to prevent any heat flow across the boundary.

In classical thermodynamics processes are hypothesized to occur in one of two possible ways, i.e., reversibly or irreversibly [21]. At the macroscopic level, however, the only kind of process that can take place is an irreversible one. Reversible processes could be carried out at this level provided they were infinitely slow, which implies they would take an infinite amount of time to complete. In practice, therefore, reversible processes are ruled out at the macroscopic level. The concept of entropy was introduced in 1854 by Rudolf Clausius in an attempt to characterize the capacity for change within a system [22, 23]. The entropy is defined only for the equilibrium states of a system, which means that all of the changes carried out in the system must be performed infinitely slowly. If this condition is satisfied, the system is deemed to remain in equilibrium at all times during the change because it passes through a continuous series of equilibrium states [24]. The change in entropy within the system is then given by the perfect differential, dS, defined as:

$$\mathrm{ds}=\delta \mathrm{q}_{\mathrm{rev}} / T \tag{1.1}$$

where δq_{rev} is the amount of heat absorbed reversibly by the system and T is the absolute temperature at which heat is absorbed. Upon integration, the equation assumes the form:

$$\Delta S = \int \delta q_{rev} / T \tag{1.2}$$

where ΔS represents the macroscopic change in the entropy of the system. The equality sign applies only in the case of perfectly reversible processes and in all other cases must be replaced by the inequality > sign. This fact informs us that the entropy will always increase for nonreversible processes.

The increase in entropy with respect to time implies that every complex system will ultimately become so degraded that all of its complexity will be destroyed. Thus, the existence of entropy points to a destructive principle of nature that operates to diminish and finally destroy complexity in the world. The question now arises whether the constructive principle of self-organization discussed earlier is compatible with the destructive principle of entropy. Is some sort of coexistence possible between these opposing principles in the real world? Perhaps somewhat unexpectedly, the laws of nature dictate that complexity will always arise in systems provided the conditions are appropriate. To understand why complexity is inevitable, we first consider some distinctions originally suggested by Prigogine [25]. Isolated systems near to equilibrium become increasingly disordered as their entropy approaches its maximum value. The only processes occurring in such systems are irreversible and they always proceed in a unique direction, namely toward maximum entropy. On the other hand, systems that are open and far from equilibrium are able to utilize the flux of matter and energy flowing through them to maintain their characteristic features. This kind of behavior is by no means restricted to living systems but is also observed in purely physical and chemical systems. In spite of the operation of what we have termed the destructive principle, however, such systems are able to sustain themselves for periods of time long in comparison to the time taken for their formation.

1.5 The Dissipative Structure

Let us now take a closer look at structures that are formed near to and far from thermodynamic equilibrium. Not only do the two types of structure differ substantially but they also arise via two quite different mechanisms.

The first type of structure, examples of which are crystals and viruses, is described as an equilibrium structure. They are so termed because they are formed in systems of low energy near to thermodynamic equilibrium. Their mechanism of formation may be characterized as a conservative or reversible kind of self-organization. In appropriate environments, these structures relax to their lowest energy state and no further input of energy is required to maintain the structure. Such structures remain highly stable for very long periods of time. Crystals are comprised of atoms or molecules bonded together in three-dimensional arrays whereas viruses consist of strands of either DNA or RNA that are surrounded by a wide variety of protein molecules that form a protective coat that includes the head structures, tail fibers and other features of the virus. Both of these systems remain remarkably stable provided that they are not disturbed in any way. In Figures 1.1–1.5 we use a number of now classic illustrations to exemplify the increasing levels of complexity that are encountered as we move from relatively simple equilibrium structures, such as those of crystals and viruses, to living systems, such as plants, animals, and the human brain.

The second of the two types of structure that we shall discuss is the so-called dissipative structure. This term was introduced by Prigogine to describe structures that dissipate matter and energy to maintain themselves [26]. All such structures are formed irreversibly in systems that are far from thermodynamic equilibrium. The dissipative process leads to a vast proliferation of structures that include geometric patterns, oscillating chemical reactions, and an extraordinary variety of living systems. The structures formed differ from equilibrium structures in that they can be maintained only if there is a sufficient flow of energy or matter through the system to support them. Dissipative structures are thus characterized by high entropy flows and universal time symmetry breaking. By the latter is meant that there exists a preferred time direction in all dissipative processes, i.e., all such processes are irreversible and tend to the equilibrium state as time approaches infinity. Before this asymptotic stability is reached, however, there is huge scope for the appearance of delicately balanced processes that are capable of sustaining extremely complex structures. Thus, in the formation of dissipative structures, what we have referred to as the constructive and destructive forces operate in concert to generate systems whose component parts may be viewed as never quite locked into place although they cohere well enough not to disintegrate into chaos. In living systems, dissipative processes are now considered to play a crucial role in that biological structures can be formed only in dissipative media and must be maintained by a continual supply of matter and energy.

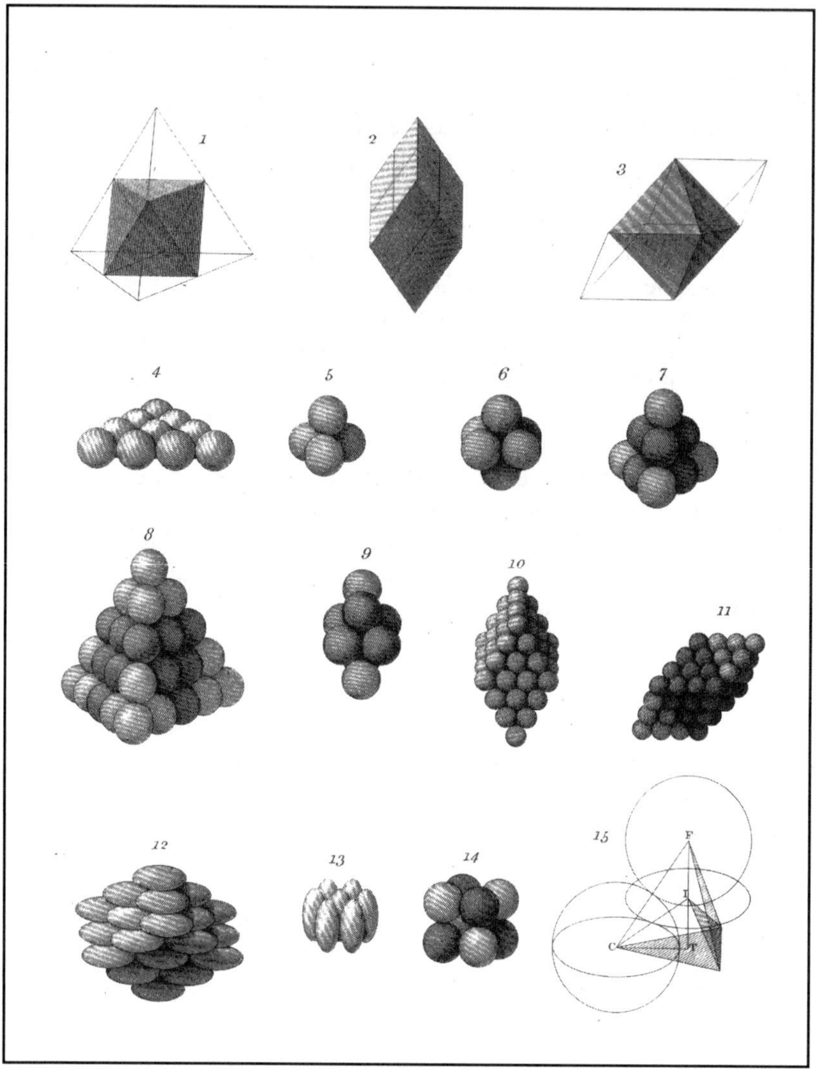

Figure 1.1: A classic illustration of crystals and their atomic structures by William Wollaston. Reproduced from *Phil. Trans. Roy. Soc.*

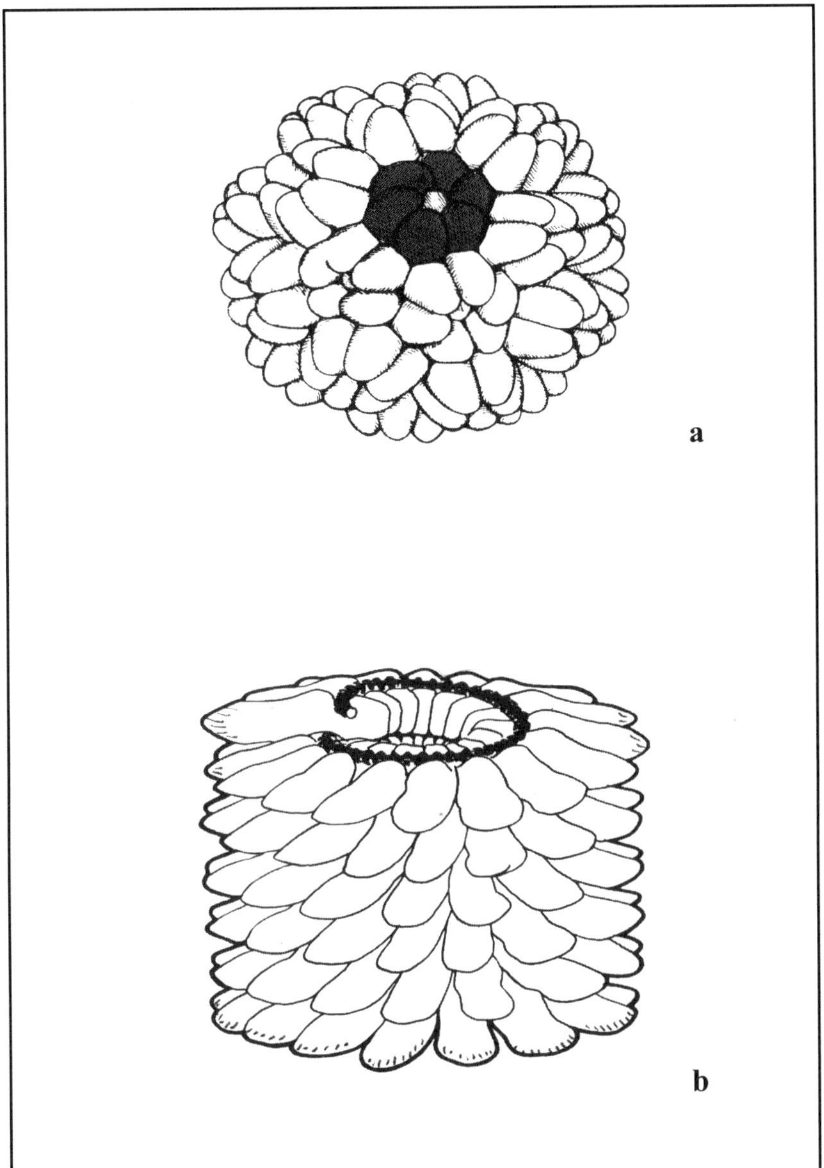

Figure 1.2: Classic illustration of models of (a) the turnip yellow mosaic virus and (b) the tobacco mosaic virus. Redrawn by the author. In (b) the axial strand of RNA is shown in addition to the protein coat.

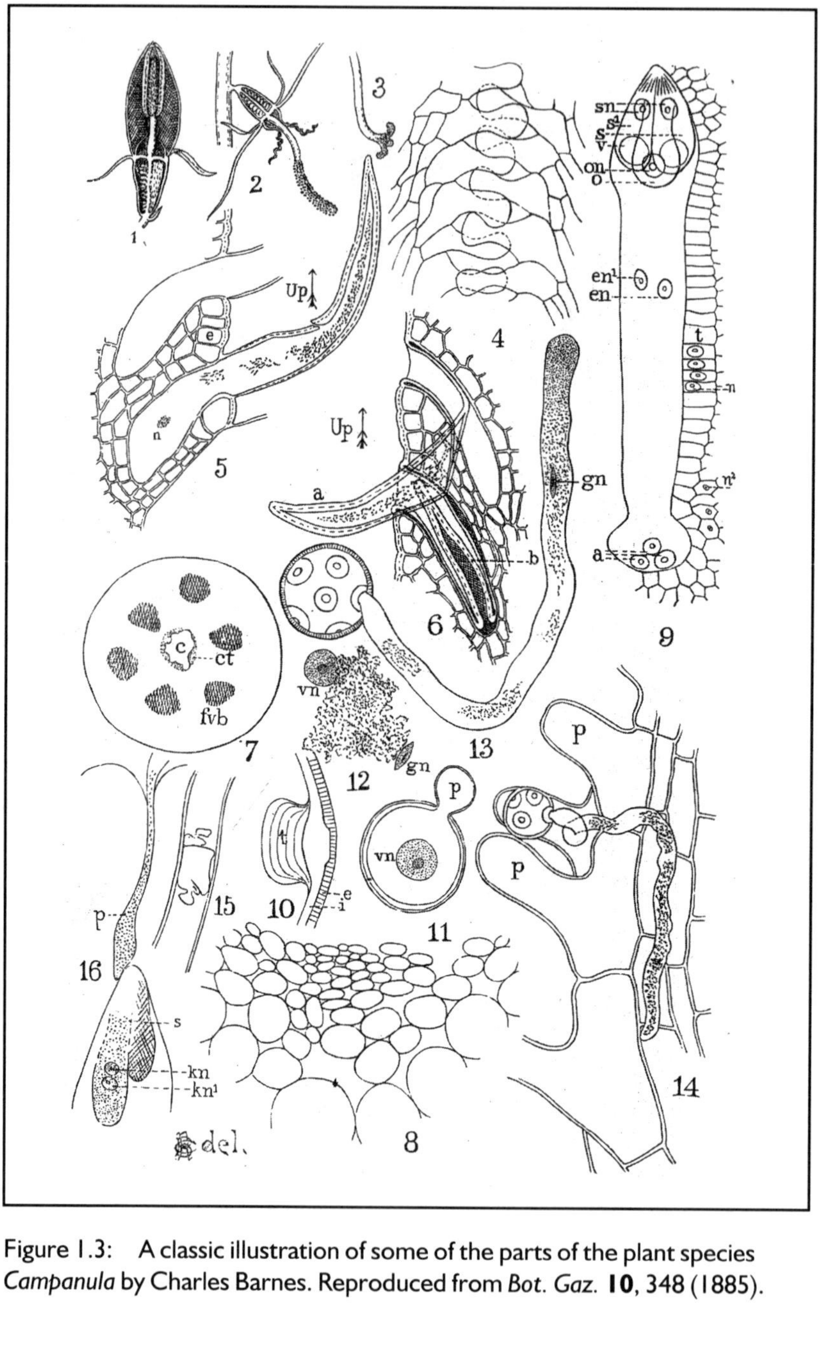

Figure 1.3: A classic illustration of some of the parts of the plant species *Campanula* by Charles Barnes. Reproduced from *Bot. Gaz.* **10**, 348 (1885).

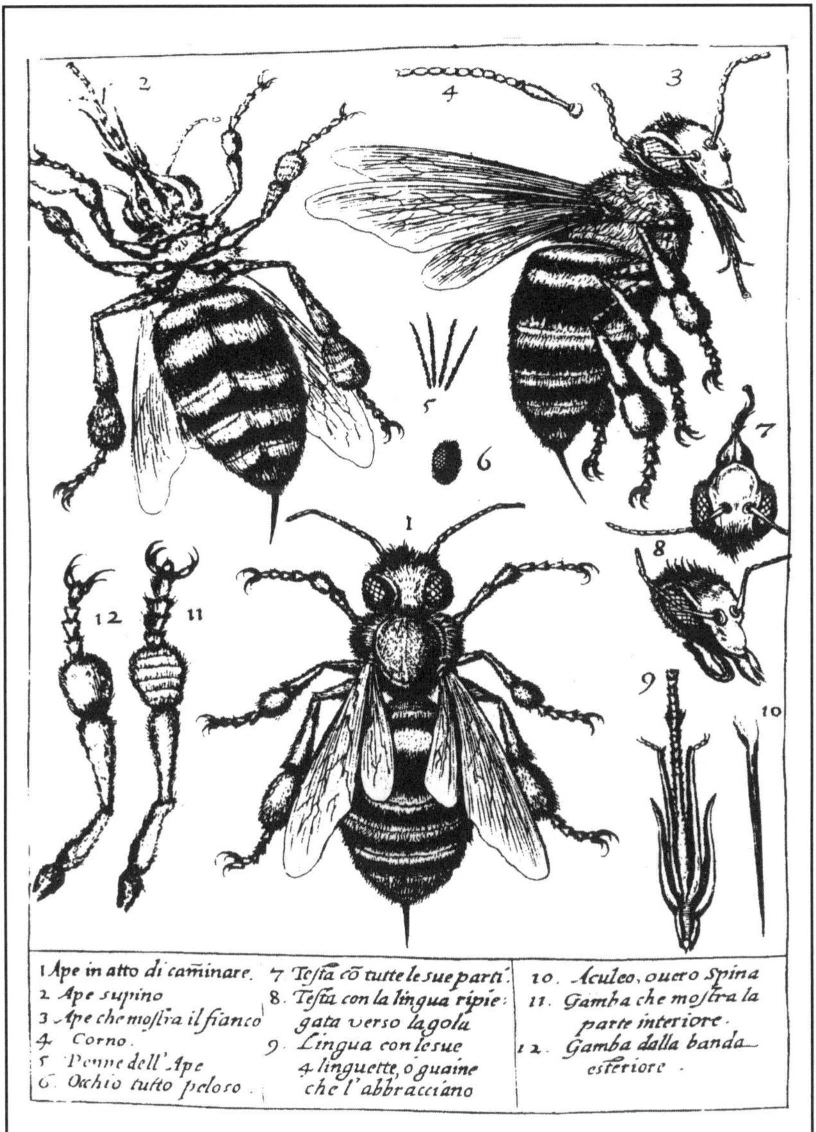

Figure 1.4: A classic illustration of bees and some of their parts by Francesco Stelluti. Reproduced from his book *Persio Tradotto* of 1630.

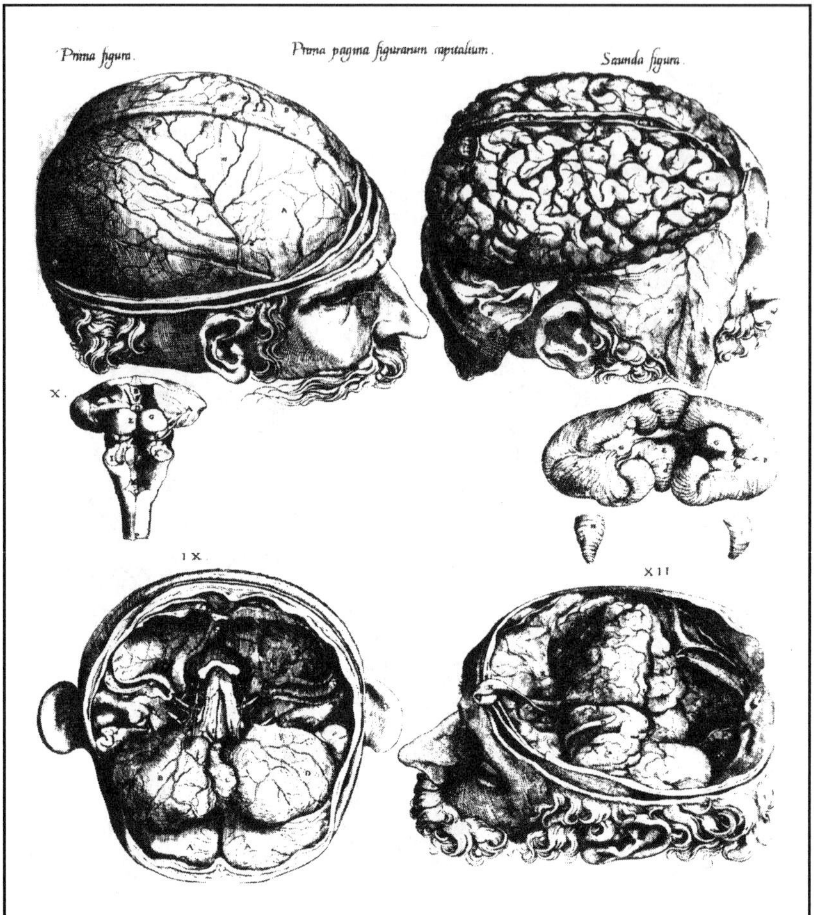

Figure 1.5: A classic illustration of sections of the human brain by Andreas Versalius from his book *Fabrica* of 1543.

It has been suggested [27] that living systems flourish near to "the edge of chaos" because in this vicinity it becomes possible for them to exploit the richest and most complex forms of behavior. Moreover, such a domain affords the best location for living systems to adapt to their environment and thereby achieve selective advantages over rival systems. The process of biological evolution is thus in general associated with an ever-increasing level of complexity in the life forms it produces. Much progress has been made in recent years in understanding how macroscopic structures are generated from interactions at the microscopic level [28]. One insight in particular has a bearing on the notion of complexity. A conceivable way of characterizing the complexity of a system is in terms of the number of degrees of freedom it has. For instance, in condensing steam to water and then freezing the water to form ice, we are effecting a dramatic reduction in the number of degrees of freedom within the system. Whereas the water molecules in steam move independently of one another without any mutual correlation of their motions, in ice the water molecules are constrained to move only minimally about their equilibrium positions in the solid lattice. By similar reasoning, the emergence of a dissipative structure may be viewed as an occurrence associated with a massive reduction in the complexity of the system in question. Thus, depending on how we interpret complexity, dissipative structures are either more or less complex than the system from which they originate.

1.6 The Information Content

Before attempting to resolve this conundrum, we shall first consider some of the ways in which the information content of a system may be described. This parameter gives us a measure of the ordering present in a system as well as the amount of usable energy contained within the system. In its origins the parameter can be traced back to the work of Boltzmann [29] on statistical thermodynamics dating from 1866. His idea was that any macroscopic system could be envisioned as consisting of a large number of particles—typically on the order 10^{23} particles. The macroscopic properties of the system would result from statistical averaging of the mechanical motions of the system's constituent particles. It was also assumed that any ensemble of the microstates of a system that yielded a given value of a macroscopic property would be equally likely to occur. On such a basis, the probability of finding the system with its known macroscopic properties is proportional to the number of microscopic states available to it, a number denoted by the symbol W. Boltzmann defined the macroscopic entropy, S, of a system as follows:

$$S = k \ln W \tag{1.3}$$

where k is the Boltzmann constant. If each of the i microscopic configurations has a differing probability, p_i, of occurring, the above equation becomes:

$$S = -k\sum_{i} p_i \ln p_i \qquad (1.4)$$

The latter equation was also derived by Boltzmann [29].

In the twentieth century there developed a growing acceptance of the existence of an intimate relationship between the concepts of entropy and information [30]. Of key importance in this context was the work of Shannon [31, 32] who conjectured that the information contained within a message must be related to its internal organization. The measure of missing information or uncertainty in messages that he proposed took the same form as the entropy defined in statistical thermodynamics. He reasoned that for a system that could exist in *n* possible states ($i = 1, \ldots, n$), each of which had an associated probability of P_i, the uncertainty in the system, U, would be given as:

$$U = \sum_{i=1}^{n} P_i \ln P_i \qquad (1.5)$$

The negative value of the parameter U is described as the information content or negentropy of the system and is usually given the symbol I. It is now widely recognized that systems operate by virtue of the fact that they contain information or, put another way, are ordered in particular ways [33]. Not surprisingly, the information content has been extensively exploited to characterize the complexity of systems. De Vree [34] has aptly summed up the situation by observing that "… intellectual information cannot just be everything we know, but … also … a matter of insight into how things hang together, … the measure of order or organization in what we seem to know."

In contrast to randomness, which can be measured by the entropy in a system, the complexity cannot be determined by one single measure that is applicable in all situations. Differing measures are needed to describe such features as the extent to which the parts of a complex system interact and the manner in which the actions of the parts are coordinated. As we have already pointed out, dissipative systems differ substantially from equilibrium systems. Against this backdrop, we now return to the issue raised at the end of Section IV, namely whether dissipative structures are more or less complex than the systems from which they originate. Dissipative systems certainly have fewer degrees of freedom than more disordered systems, but this is not the relevant consideration in the present context. Because dissipative systems operate at comparatively low entropies, they may be regarded as storage devices for negentropy. This fact alone, however, tells us nothing about the potential of the system to develop into a more complex and, possibly, living system. A new measure is thus called for that can provide an indication of the potential within the system to become more complex. Such a measure was recently introduced by Korzeniewski [35], who made use of the fact that the purposefulness of living organisms is dependent upon a network of regulatory mechanisms, i.e., negative feedback loops, which enable organisms to survive and reproduce. His expression for the amount of purposeful biological information, I_B, within a system is:

$$I_B = \log_2 (N.2) \tag{1.6}$$

where I_B is stated in bits, N is the number of feedback loops in the organism, and 2 is the minimum number of states a negative feedback loop can distinguish. An important advantage of measures of this kind is that dissipative systems can indeed be shown to be more complex than disorganized systems.

1.7 The Complex System

Because, in the words of Paley [26], the complex systems of the natural world "surpass the contrivances of art, in complexity, subtilty, and curiosity of the mechanism," we pause at this point to take a closer look at some of these systems and how they are formed. It seems appropriate to start with atoms and small molecules, which means that quantum-mechanical considerations will be of prime importance. In passing, however, we would just cite the comment of Peres and Zurek [37] that "[t]here is nothing in quantum theory making it applicable to three atoms and inapplicable to 10^{23}." Studies of microscopic complex systems that exploit information-theoretical ideas have come to the fore only in the last decade [38]. Consequently, there is a need for much further work in this area. To date, approaches have been made that have included determination of the quantal information entropies for atoms [38]; usage of the information entropy as a measure of the quality of approximate atomic wave functions [39]; analysis of maximum entropy electron momentum densities in diatomic molecules [40]; development of a method of calculating the many-body correlational electron entropies in molecular species [41]; and investigation of the foundations of quantum mechanics [42]. A number of larger systems, such as quasi-periodic structures [43] and blackbody radiation emitters [44], have been similarly modeled. The possibility of adopting information measures other than those based on Shannon's formulation has also been discussed [45].

It is now widely accepted that biological cells evolved from crudely replicating molecules some 3.8 billion years ago [46]. The earliest kinds of cells were procaryotic, meaning that the cells had no nucleus and that their DNA was not separated from the rest of the cell. Perhaps the most important event in the entire history of the evolution of life on earth occurred some 1.9 billion years ago when the first eucaryotic cells were formed, possibly from the fusion of procaryotic cells. Eucaryotic cells not only have a nucleus but their DNA is enclosed within a membrane inside the cell. This development was of major significance because all complex life forms are based on eucaryotic cells whereas procaryotic life forms are much simpler and consist primarily of bacteria. Cells are built up of many millions of molecules that interact in manifold ways to yield an integrated and extremely complex unit. All of the cellular molecules must work together in a precisely regulated modus operandi to render possible the maintenance of life functions. Indeed, the inner workings of a cell have been linked to the coordinated working of a fully automated chemical factory [47]. The activities of a cell involve metabolic cycles or regulatory mechanisms that are ultimately controlled by information contained within the cell's DNA. Eigen [48] has shown that the amount of information that can be reliably transferred to a new cell during reproduction is of critical importance in evolutionary processes. By means of sophisticated techniques (that involve essentially "proof-reading" the new information), reproduction by eucaryotic cells can raise the accuracy threshold to over 10^9 nucleotides—which is about the amount of information required to encode the human genome [49].

As mentioned in Section 1.3, the human brain represents evolution's acme of complexity. In Figure 1.6 we show in schematic form some of the more important stages in the formation of a functioning human brain. In its workings, the brain is the epitome of holistic systems: not only must the brain be considered as a whole but as we progress through the stages shown in Figure 1.6, we encounter to an ever-increasing extent emergent effects which cannot be understood solely in terms of the characteristics of the constituent parts. For instance, although individual neurons are not capable of perceiving the world, this facility becomes a reality when they are integrated into appropriate neural networks [50]. Most of the neurons in the central area of the cerebral cortex possess voltage-dependent

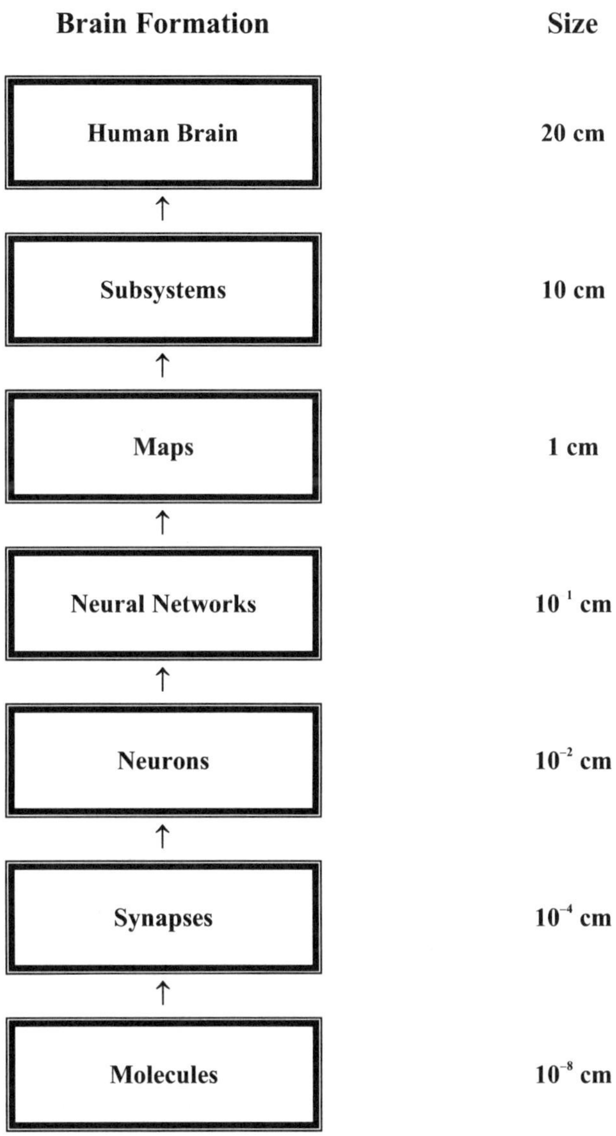

Figure 1.6: A schematic illustration of some of the major subsystems in the structure of the human brain. Based on reference [28].

conductances that confer very complex electroresponsiveness on the cells [51]. Examples include digital spikes of electrical activity and sustained chemical oscillations. The spikes are an integral part of the firing patterns in cerebral neural networks whereas the oscillations become synchronized and lead to coherent oscillatory behavior at various frequencies in the cortical areas [52]. Thus, a variety of complex and oscillatory phenomena are generated that are reminiscent of spatiotemporal deterministic chaos in physical systems [53]. Each of the neural states of consciousness will have a high information content because all such states are brought into being as the outcome of a selection process. Clearly, the selection of a given conscious state from a virtually infinite number of theoretically possible states results in a major reduction in the uncertainty of the system. The brain thus manifests two quite different kinds of complexity, viz. (i) structural complexity which relates to the way in which its neurons are interconnected, and (ii) neural complexity which follows from the ability of the neural networks to generate conscious states with a high information content.

1.8 The Defining Moment

Because of the fuzzy nature of the concept and the fact of its multifarious usage in a plethora of different situations, any attempt to define the meaning of the word complexity is a remarkably challenging, if not futile, exercise. In spite of some grave reservations, however, we now attempt to offer an idea of how the word has come to be used in the physical, chemical and biological sciences. The word complexity itself derives from two Latin words: *com*, meaning together, and *plectere*, meaning to braid or entwine. In its original extended meaning the word therefore signified embracing, encompassing or inclusion. In its modern meaning, especially in scientific contexts, complexity is a term that provides us with a conceptual and mathematical way of getting to grips with systems containing multiple and interdependent parts or subsystems. Some of the earliest attempts to arrive at a useful working definition were made by Pierce [54], Weaver [55], and Simon [56]. Examples of more recent attempts include those of Pippenger [57], Yates [58], Klir [59], Grassberger [60], Wright [61], and Goldenfeld and Kadanoff [62]. Perhaps not surprisingly, there is substantial divergence in the ideas and opinions expressed by these authors. In the circumstances, rather than examine the work of many authors, it seems more appropriate to focus our attention on one specific author and we have arbitrarily selected the work of Rescher [63]. His work contains what we consider to be a number of compelling ideas. Rescher recognizes that complexity itself is a complex notion and that systems can be more or less complex in quite different ways. He also maintains that the concept of complexity "fuses and integrates a plurality of distinct elements into an elaborately articulated coordination" [63, p. 10]. We present in our Table 1.1 a listing of the major categories of complexity according to Rescher's perspective.

All of the categories of complexity we have listed have been employed in recent decades in the physical, chemical and biological sciences. We shall make no further mention of the broad area of irreversible thermodynamics, the production of negentropy, and the emergence of dissipative structures. This area we feel has been sufficiently covered already in our overall review. Neither will we make reference to the closely related area of chemical pattern formation and oscillating chemical systems usually conflated nowadays under the generic heading of chemical chaos [64, 65]. Some comment is called for, however, on the elaboration of complexity measures for individual chemical species. Both formulaic and structural complexity measures have been introduced and much effort has been expended on devising measures that reliably reflect underlying chemical realities. The generally adopted approach involves the initial representation of the chemical species by means of a chemical graph [66] and this is followed by some algorithmic procedure that is carried out on the graph. Examples of the measures formulated in this way include Rashevsky's information content of a graph [67], Bertz's complexity of a graph [68], and the topological complexity indices of Bonchev [69]. These and related measures are comprehensively discussed in the succeeding three chapters. Complexity measures have also been developed for atoms [70, 71], macromolecular knots [72], and hydrogen-bonded networks in proteins [73]. Algorithmic complexity measures are based on the length (in bits) of the shortest computer program capable of yielding a given binary string of symbols [74]. Such measures have been adopted in the characterization of proteins [75, 76] and snowflakes [77].

When we move toward the biological sciences, the realm in which virtually the entire populace consists of living systems, a difficult task confronts us when we endeavor to apply any of the definitions we have developed hitherto. The great variety of living systems, the remarkable sophistication of their actions, and their legendary adaptability would seem to render our somewhat ingenuous definitions inadequate or even inappropriate. We have already pointed out in the case of the human brain that this system is comprised of numerous subsystems, all of which interact

FORMULAIC COMPLEXITY
Descriptive Complexity: The length of the system description
Algorithmic Complexity: The length of the system instructions
Computational Complexity: The time required to solve a problem
COMPOSITIONAL COMPLEXITY
Constitutional Complexity: Number of parts or components
Taxonomical Complexity: Number of different kinds of parts
STRUCTURAL COMPLEXITY
Organizational Complexity: Number of permutations of parts
Hierarchical Complexity: Elaborateness of subsystem relationships
FUNCTIONAL COMPLEXITY
Operational Complexity: Variety of modes of operation
Nomic Complexity: Degree of elaboration of the governing laws

Table 1.1. A listing of the principal categories of complexity based on the formulations of Rescher [63].

with one another to form an integrated and holistic unity. Moreover, biological systems generally need to be characterized in terms of concepts such as the purposeful biological information content that was defined in Equation (1.6). It would seem that we are currently only at the threshold of developing suitable concepts and that several more will likely be necessary. In spite of these widely recognized problems, however, several workers have attempted to model evolutionary processes by simulated computer techniques. Thus, Sendhoff et al. [78] sought to delineate the differing kinds of information that were involved in the evolutionary process as a whole. The four kinds that they identified were (i) the entropy, which represented the information gathered during the evolutionary process at the lowest level; (ii) the algorithmic complexity, which represented the coded information content in the system without reference to a statistical ensemble; (iii) the information of the phenotype, which related to the potential and actual ability of systems to process information and therefore to their adaptability; and (iv) the self-organized information acquisition, which pertained to the generation of information within the system as a result of its anticipation of future events. Schemes such as this can certainly make no claim to elucidate fully the systems they model. Indeed, a detailed understanding of the entire process of evolution is probably decades away.

1.9 The Concluding Word

Although our primary intention in this chapter has been to introduce the subject of complexity and, in particular, to focus on its occurrence and role in the physical, chemical and biological sciences, we wish to stress that complexity is ubiquitous. By this we mean that complexity is to be found in all areas of our world and in all aspects of our existence. The concepts of complexity that have been developed over the past half century are so eclectic that they are now accepted to cut across specialisms and apply to most disciplines. In this regard, complexity studies have not only opened up whole new vistas in our thinking but they have also had the gratifying result of establishing important links between the sciences and engineering on the one hand and the arts and humanities on the other hand. Complexity in all its ramifications thus represents an important bridging discipline between the arts and the sciences. Consider, for instance, studies on the human brain—a field of endeavor of ever-increasing prominence. As the most complex of all the machines in the world, the human brain has been interpreted and elucidated by workers in disciplines as diverse as biology, chemistry, physics, computer science, anatomy, physiology, psychology and sociology. Even so, it is fair to comment that the functional principles of the brain are still largely enigmatic. It seems likely that much of the twenty-first century will be devoted to studies that have as their ultimate goal the understanding of highly complex systems such as the human brain.

A key question that needs to be asked is whether it will be possible to understand the brain using the traditional tools of the physical sciences. These sciences adopt a relentlessly reductionistic approach to all kinds of complexity and always attempt to reduce complex systems to irreducible essentials. But the brain and biological systems in general may not yield to such a line of attack. The biological sciences emerged at the end of the nineteenth century with all the necessary prerequisites to develop into fully fledged sciences. Biological systems had been collected, sampled, described, classified and categorized and it was only in the twentieth century—especially in the latter half—that serious attempts were made to reduce biological phenomena to their underlying physics and chemistry.

What twentieth-century work has revealed, however, is that complex biological systems cannot be reduced to much simpler physical systems! The two major reasons for this are that the physical systems themselves are not simple and for both biological and physical systems the approach that seems closest to reality is one based on the notion of holism. This notion is sometimes encapsulated in the observation that a system is more than the sum of its parts. Merely examining the neurons in the brain isolation would never tell us anything about integrated and holistic concepts such as perception and intelligence. Not surprisingly, in recent decades there has been an inexorable move away from reductionist thinking and toward holistic interpretations of nature. Even something as seemingly simple as an atom is now commonly interpreted in terms of quantum theory—a holistic theory and the first such theory to be put forward in the physical sciences. Taking matters yet further in this direction is the now widely accepted idea that the holistic concept of intelligence, which has been defined as the ability of a dissipative system to store negentropy within its structure, need not be restricted to living systems. All systems can now considered to have some measure of intelligence according to this definition. It would seem that the words of Nietzsche in our opening quote were remarkably prescient: What is real … is neither simple nor is it reducible to simplicity.

1.10 References

1. W.H.D. Rouse (Ed.), *St. Augustine's Confessions, with an English translation by William Watts (1631)*, Vol. II, Book XI. 14, Loeb Classical Library, Harvard University Press, Cambridge, MA, 1912, p. 239.
2. P. Holland, *The Quantum Theory of Motion: An Account of the de Broglie-Bohm Causal Interpretation of Quantum Mechanics*, Cambridge University Press, Cambridge, UK, 1993, p. 26.

3. D.H. Rouvray, in: *Concepts in Chemistry: A Contemporary Challenge*, D.H. Rouvray (Ed.), Research Studies Press, Taunton, UK, 1997, pp. 5–7.
4. D.L. Medin, *Amer. Psychol.* **44**, 1469 (1989).
5. K.F. Johansen, *A History of Ancient Philosophy*, transl. H. Rosenmeier, Routledge, London, 1998, pp. 47–54; 218–220.
6. H. Primas, in *Inside Versus Outside*, H. Atmanspacher and G.J. Dalenoort (Eds), Springer, Berlin, 1994, p. 163.
7. A. Einstein, B. Podolsky and N. Rosen, *Phys. Rev.* **47**, 777 (1935).
8. D. Bohm and B.J. Hiley, *The Undivided Universe: An Ontological Interpretation of Quantum Theory*, Routledge, London, 1993, p. 58.
9. E. Schrödinger, *Ann. Phys.* **79**, 361, 489, 7334 (1926).
10. D. Bohm and B.J. Hiley, *The Undivided Universe: An Ontological Interpretation of Quantum Theory*, Routledge, London, 1993, p. 59.
11. J. Halliwell and S.W. Hawking, *Phys. Rev.* **D31**, 1777 (1985).
12. P. Davies, *The Cosmic Blueprint*, Simon and Schuster, New York, 1988, p. 5.
13. K. Popper and J. Eccles, *The Self and its Brain*, Springer, Berlin, 1977, p. 61.
14. P. Davies, *The Cosmic Blueprint*, Simon and Schuster, New York, 1988, p. 200.
15. L. Bernstein, in *The Scientific American Book of the Cosmos*, D.H. Levy (Ed.), St Martin's Press, New York, 2000, p. 65.
16. J. Gribbin, *Companion to the Cosmos*, Weidenfeld and Nicholson, London, 1996, pp. 162–163.
17. S. Kauffman, *At Home in the Universe: The Search for Laws of Self-Organization and Complexity*, Oxford University Press, New York, 1995, pp. 43–45.
18. A. Smith, *The Mind*, Hodder and Stoughton, London, 1984, chap. 1, reprinted in J. Carey (Ed.), *The Faber Book of Science*, Faber and Faber, London, 1995, pp. 440–444.
19. G. Tonini and G.M. Edelman, *Science* **282**, 1846 (1998).
20. D. Hull, in *Entropy, Information, and Evolution*, B. Weber, D. Depew and J. Smith, (Eds), MIT Press, Boston, MA, 1988, p. 3.
21. M. Planck, *Ann. Phys.* **30**, 563 (1887).
22. R. Clausius, *Ann. Phys.* [2], **93**, 481 (1854).
23. S.G. Brush, *Arch. Hist. Exact Sci.* **12**, 1 (1974).
24. P. Coveney and R. Highfield, *The Arrow of Time*, Allen, London, 1990, p. 155.

25. I. Prigogine, *Introduction to Thermodynamics of Irreversible Processes*, 2nd edn, Interscience, New York, 1961, p. 93.
26. I. Prigogine and G. Nicolis, *J. Chem. Phys.* **46**, 3542 (1967).
27. C.G. Langton, *Physica*, **D42**, 12 (1990).
28. K. Mainzer, *Thinking in Complexity: The Complex Dynamics of Matter, Mind, and Mankind*, 3rd edn, Springer, Berlin, 1997, pp. 91–106.
29. L. Boltzmann, *Wien. Ber.* **53**, 195 (1866).
30. R.U. Ayres, in: *Organization and Change in Complex Systems*, M. Alonso (Ed.), Paragon House, New York, 1990, pp. 207–208.
31. C.E. Shannon, *Bell System Technical J.* **27**, 379 (1948).
32. C.E. Shannon and W. Weaver, *The Mathematical Theory of Communication*, University of Illinois Press, Urban, IL, 1949.
33. T. Stonier, *Information and the Structure of the Universe: An Exploration into Information Physics*, Springer, London, 1990, pp. 54–57.
34. J.K. De Vree, *BioSystems*, **38**, 221 (1996).
35. B. Korzeniewski, *J. Theor. Biol.* **209**, 275 (2001).
36. W. Paley, *Natural Theology*, Knight, London, 1802, p. 16.
37. A. Peres and W.H. Zurek, *Am. J. Phys.* **50**, 807 (1982).
38. A. Bhattacharya, B. Talukdar, U. Roy and A. Ghosh, *Int. J. Theor. Phys.* **37**, 1667 (1998).
39. A. Nagy and R.G. Parr, *Int. J. Quantum Chem.* **58**, 323 (1996).
40. J. Antolin, A. Zarzo, J.C. Angulo and J.C. Cuchi, *Int. J. Quantum Chem.* **61**, 77 (1997).
41. P. Gersdorf, W. John, J.P. Perdew and P. Ziesche, *Int. J. Quantum Chem.* **61**, 935 (1997).
42. B.B. Kadomstev, *Uspekhi Khim.* **37**, 425 (1994).
43. E. Maciá and F. Dominguez-Adame, *Phys. Rev.* **E50**, R679 (1994).
44. J. Fort, J.A. Gonzalez, J.E. Llebot and J. Saurina, *Contemp. Phys.* **40**, 57 (1999).
45. W.H. Zurek, *Phys. Rev.* **A40**, 4731 (1989).
46. M. Ridley, *Mendel's Demon*, Weidenfeld and Nicholson, London, 2000, p. 13.
47. B.-O. Küppers, *Information and the Origin of Life*, MIT Press, Cambridge, MA, 1990, p. 8.
48. M. Eigen, *Naturwiss.* **58**, 465 (1971).
49. B.-O. Küppers, *Molecular Theory of Evolution*, transl. P. Woolley, Springer, Berlin, 1985, p. 151.
50. W. Singer, *Prog. Brain Res.* **95**, 461 (1993).
51. R.R. Llinas, *Science*, **242**, 1654 (1988).

52. M. Steriade, D.A. McCormick and T.J. Sejnowski, *Science*, **262**, 679 (1993).
53. A. Destexhe, *Phys. Rev.* **E50**, 1594 (1994).
54. C.S. Peirce, *Collected Papers*, Vol. 6, Harvard University Press, Cambridge, MA, 1935, pp. 39–41.
55. W. Weaver, *Amer. Sci.* **36**, 536 (1948).
56. H.A. Simon, *Proc. Am. Phil. Soc.* **106**, 467 (1962).
57. N. Pippenger, *Sci. Am.* **238**, 114 (1978) [June].
58. F.E. Yates, *Am. J. Physiol.* **235**, R201 (1978).
59. G.J. Klir, *Sys. Res.* **2**, 131 (1985).
60. P. Grassberger, *Int. J. Theor. Phys.* **25**, 907 (1986).
61. R. Wright, *Three Scientists and their Gods: Looking for Meaning in an Age of Information*, Times Books, New York, 1988, chap. 18, pp. 204–211.
62. N. Goldenfeld and L.P. Kadanoff, *Science*, **284**, 87 (1999).
63. N. Rescher, *Complexity: A Philosophical Overview*, Transaction Publishers, New Brunswick, NJ, 1998, chap. 1, pp. 1–24.
64. S.K. Scott, *Chemical Chaos*, Clarendon Press, Oxford, 1991.
65. S.K. Scott, *Oscillations, Waves and Chaos in Chemical Kinetics*, Oxford University Press, London, 1994.
66. D.H. Rouvray, in: *Chemical Graph Theory: Introduction and Fundamentals,* Abacus/Gordon and Breach, London, 1991, chap. 1, pp. 1–16.
67. N. Rashevsky, *Bull. Math. Biophys.* **30**, 229 (1955).
68. S.H. Bertz, *J. Am. Chem. Soc.* **103**, 3599 (1981).
69. D. Bonchev, *SAR QSAR Environ. Res.* **7**, 23 (1997).
70. D. Bonchev and V. Kamenska, *Croat. Chem. Acta*, **51**, 19 (1978).
71. J.B. Hendrickson, P. Huang and A.G. Toczko, *J. Chem. Inf. Comp. Sci.* **27**, 63 (1987).
72. D. Sumners, *Computers Chem.* **14**, 275 (1990).
73. G.A. Ateca and X. Luo, *J. Mol. Struct. (Theochem)*, **501**, 479 (2000).
74. W.H. Zurek, *Nature*, **341**, 119 (1989).
75. T.G. Dewey, *Phys, Rev.* **E54**, R39 (1996).
76. O. Weiss, M.A. Jimenez-Montano and H. Herzel, *J. Theor. Biol.* **206**, 379 (2000).
77. F. Papentin, *Naturwiss.* **67**, 174 (1980).
78. B. Sendhoff, C. Pötter and W. von Seelen, in: *Unifying Themes in Complex Systems*, Y. Bar-Yam (Ed.), Perseus Books, Cambridge, MA, 2000, chap. 41, pp. 453–472.

Chapter 2

ON MOLECULAR COMPLEXITY INDICES

S. Nikolić and N. Trinajstić
The Rugjer Bošković Institute, P.O.B. 180,
HR-10002 Zagreb, The Republic of Croatia

I.M. Tolić
Harvard School of Public Health – Physiology Program,
Harvard University, 665 Huntington Avenue, Boston, MA 02115, USA
(Permanent address: The Rugjer Bošković Institute, P.O.B. 180,
HR-10002 Zagreb, The Republic of Croatia)

G. Rücker and C. Rücker
Department of Mathematics, University of Bayreuth,
D-95440 Bayreuth, Federal Republic of Germany

Measuring complexity in chemical systems, biological organisms or even poetry requires the counting of things.

S.H. Bertz and W.F. Wright [1]

2.1 Introduction

The concept of complexity is rather difficult to define [2]. Some authors have even stated that nobody knows how to define complexity [3]. This did not prevent the science publishing company John Wiley and Sons, Inc. of New York from launching in 1995 an international journal entitled "*Complexity*" (Harold Morowitz, Editor-in-Chief; John L. Casti, Executive Editor) with the aim of providing "global coverage of all aspects of the science of complexity in a readable, topical and informative way."

The manifestations of complexity can be found everywhere in nature and in all kinds of human activities [2,4–7]. Different aspects of complexity are encountered in mathematics (see, for example, a paper by Mowshowitz [8] on the complexity of graphs), physics (see, for example, a note by Kerr [9] on the complexity of the Uranian magnetic field), chemistry and biochemistry (see, for example, a paper by Papentin [10] on the complexity of chemical and biochemical structures), biology (see, for example, the discussion in a paper by Hinegardner and Engelberg [11]). There is also a subdiscipline of mathematics called complexity theory which is concerned with the efficiency of algorithms and the hardness of computational problems [12,13]. Therefore, it is no wonder that this concept and the related theory, complexity theory, are discussed even in science fiction [14], where they are called by the author, Michael Crichton, "the science of the twenty-first century." Some researchers have used complexity for studying problems as diverse as the complexity of poems [15] and the origin of life [16]. Hence, it is not surprising that there are many attempts in the scientific literature to define complexity and even to quantify it [17,18]. One certainly needs a quantitative measure of complexity in order to compare and classify complex systems.

The perception of complexity is in fact subjective. Consider the streets of a city. For the citizens to find their way through the streets is as simple as can be, but a stranger may perceive them as a terribly complex maze. Known things may appear simple, unknown ones complex. The same is valid for the learning process: Things we know already are simple, things we do not understand are complex.

The word *complexity* is made up from the Latin roots: *com* (meaning *together*) and *plectere* (meaning *to plait*). In Webster's Ninth New Collegiate Dictionary [19] the term *complexity* is defined as 1: *the quality or state of being complex,* 2: *something complex*, while the term *complex* is defined as *a whole made up of complicated or interrelated parts*. Similar definitions can be found in other dictionaries [20].

Our understanding of complexity can be summarized as follows. A system is complex if it consists of a number of components interacting with each other in many different ways, so that these interactions sometimes lead to unexpected collective properties (fashionably called *emergent properties* [2]). Thus, increasing complexity is associated with an increasing number of components in a system and with increasing versatility of their interactions.

A complex system can be represented by a graph, with the graph's vertices representing the components and the edges representing the interactions. On the other hand, chemists use graphs to represent molecules (with vertices and edges standing for atoms and bonds, respectively) or reaction networks (with vertices and edges representing compounds and reactions, respectively). Therefore, the methods developed in chemistry to quantify the complexity of molecules or reaction networks are in fact useful for quantifying the complexity of graphs and thus of systems in general. "Molecular complexity indices" should therefore be applicable far beyond chemistry.

Nevertheless, in this article we are concerned mainly with the complexity of molecules, though we use (as did earlier authors) chemical and graph-theoretical terms interchangeably [21–24]. As an example of a molecular graph we show in Figure 2.1 the hydrogen-suppressed branched tree corresponding to 2-methylbutane. A tree is a connected acyclic graph. A graph is acyclic if it has no cycles.

Since complexity is a difficult concept to define and since people have an intuitive feeling about it, complexity appears to be a fuzzy concept. However, there are many fuzzy concepts in chemistry that are nevertheless very useful [25]. The first such fuzzy concept that comes to mind, which has been permeating chemistry for more than a century, is aromaticity. The fuzziness of this concept did not prevent chemists from attempting to quantify aromaticity [26]. The situation is similar with complexity. Chemists freely use the terms *complex* and *complexity* in describing molecules and their reactions [27]. Therefore, it is not unexpected that contemporary chemical literature contains a number of attempts to define and to quantify the complexity of molecules and reaction paths; these are given here in roughly chronological order [1,28–63].

Yet, one should be aware that several levels of molecular complexity may be distinguished, as of complexity in general [37]. We will discuss here a number of complexity indices related to the so-called structural or topological or graph-theoretical complexity, which is determined completely by the particular adjacency of a graph's vertices, i.e., issues of metrics and geometry will not be dealt with.

(a) 2-methylbutane

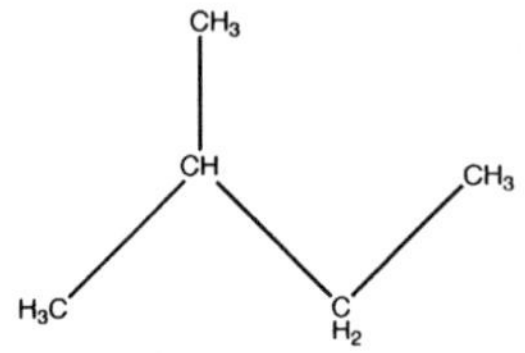

(b) Tree corresponding to 2-methylbutane

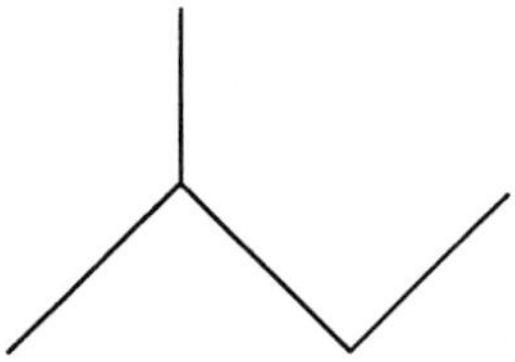

T

Figure 2.1: Hydrogen-suppressed tree T representing 2-methylbutane.

Trees representing *n*-hexane (T_1) and 2-methylpentane (T_2)

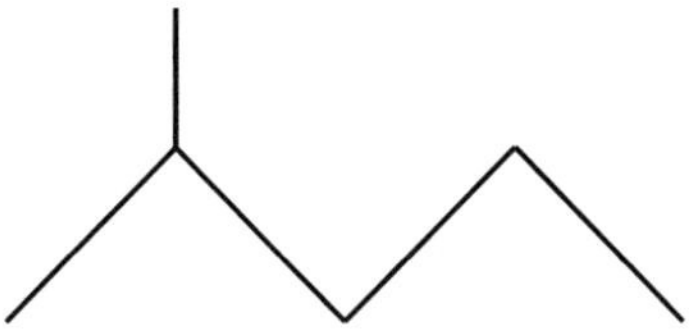

T_1 T_2

Figure 2.2: Which of the two trees is more complex?

Cycles representing cyclobutane (G_1) and cycloheptane (G_2)

G_1 G_2

Figure 2.3: Which of the two cycles is more complex?

The so-called elemental or compositional complexity, which is determined by the partition of a graph's vertices into classes of different types (atoms of different elements, heteroatoms), though easy to perceive and dealt with early in chemical history (see references given in [42, 63(b)]), will not be discussed separately here but rather under the heading of heteroatoms.

The so-called dynamical complexity, which is perceived by some authors as an additional complexity feature in fluctuating molecules such as bullvalene [64], in fact enhances a molecule's symmetry (and therefore, according to most authors, lowers its complexity). It will therefore be considered implicitly under the heading of symmetry, rather than receive separate treatment.

Complexity indices arose from the need to compare molecules according to their structural complexity. For example, which of the two trees, representing two isomeric hexanes, in Figure 2.2, is more complex? Qualitatively we can give an answer relying on our intuition: The tree T_2 is more complex than the tree T_1 since T_2 is a branched tree.

If we wish to compare two simple cycles such as those given in Figure 2.3, the intuitive complexity criterion is the size of a cycle. Therefore, in the example considered, the larger cycle is the more complex structure. Examples considered in Figures 2.2 and 2.3, however, could be resolved much more coherently if some numerical criterion is used instead of intuitive reasoning.

This chapter is structured as follows. In the next section we discuss criteria for molecular complexity indices. In the third section we briefly review a number of molecular complexity indices. The fourth section contains a comparison between various complexity indices for several fundamental classes of molecules.

2.2 Criteria for Complexity Indices

Apparently a few papers exist discussing or at least stating criteria for graph complexity indices [1]. A first catalogue of criteria is due to Minoli, 1975 [65]: A complexity index should …

Mi1: … monotonically increase with the number of vertices.
Mi2: … monotonically increase with the number of edges.
Mi3: … reflect the degree of connectedness.
Mi4: … satisfy our intuitive feel for complexity, assigning a high value to a graph which looks complicated, and vice versa.

Bonchev and Polansky in 1987 [37] discussed the following criteria: A complexity index should …

BoP1: … be independent of the nature of the system.
BoP2: … be specified within a unique theoretical concept.
BoP3: … take into account different complexity levels and their hierarchy.
BoP4: … exhibit a stronger dependence on the relations between the elements of a system than on the number of elements.
BoP5: … increase monotonically with the number of different complexity features.
BoP6: … agree with the intuitive idea of complexity.
BoP7: … differentiate non-isomorphic systems.
BoP8: … not be too sophisticated.
BoP9: … be applicable to practical purposes.

Bertz and Wright in 1998 [1] completely rejected intuition, focused on logic and came up with what can be paraphrased as follows: A complexity index should …

BeW1: …increase monotonically with factors that increase complexity as embodied in homologous series.
BeW2: …increase with increasing branching and cyclicity.
BeW3: …increase with the number of multiple edges.
BeW4: …take into account symmetry as a simplifying feature. (But this is "not always desirable.")
BeW5: …be sensitive to heteroatoms.

Rücker and Rücker recently [58] demonstrated the desirable behavior of their complexity indices along the following lines: A complexity index should increase with …

RR1: … increasing graph size.
RR2: … increasing branching.
RR3: … increasing cyclicity.
RR4: … increasing number of multiple edges.
RR5: … increasing number of weighted vertices (heteroatoms).
RR6: … decreasing symmetry (optional).

Bertz and Zamfirescu [61] after repeating the homology criterion (BeZ1) formalized criteria BeW2, RR2,3 by stating that a measure of intrinsic complexity must obey the inequalities

BeZ2: $I(K_n) \geq I(C_n) \geq I(P_n)$

and

BeZ3: $I(K_n) \geq I(K_{1,n-1}) \geq I(P_n)$

for all n, where K_n, C_n, P_n, and $K_{1,\,n-1}$ are the complete graph, the cycle graph, the path graph and the star graph of n vertices, respectively.

Of those criteria that are mentioned only once Mi3 is trivially satisfied as long as we deal with connected graphs, BoP7/BoP8 contradict one another if taken literally, BoP1–3,9 are of a very general, non-numerical nature. Symmetry as a phenomenon to be accounted for is mentioned twice (BeW4, RR6). We discuss the role of symmetry below.

Intuition as a guiding principle appears twice (Mi4, BoP6), while BeW sacrifice it in favor of logic. In our opinion intuition nevertheless can provide (here as everywhere in science) useful guidelines, at least a counterintuitive result has to be evaluated critically and may provide a clue to deeper insight. Further, it seems difficult to lay the foundation of a discipline without intuition; the choice of the criteria included in such a catalogue is a matter of intuition in itself.

All authors agree that a (molecular) complexity index should satisfy the principle of homology (Mi1, BoP4,5, BeW1, RR1, BeZ1), and reflect structural features such as branching, cyclicity, multiple edges, heteroatoms (Mi2, BoP5, BeW2,3,5, RR2–5, BeZ2,3). It is primarily according to these criteria that we shall evaluate some complexity indices in what follows.

2.3 Survey of Several Molecular Complexity Indices

We now review a number of the more recent molecular complexity indices. For the prehistory of molecular and graph complexity see references [42, 53, 56, 63(b)] and the references cited therein. It is quite possible that among the numerous topological indices constructed hitherto there are another few whose ability to serve as a complexity index has not yet been recognized, as happened with the Zagreb indices [1].

For each index covered we give its definition, a figure illustrating its calculation for a simple example, closed formulas for the series of *n*-alkanes and unbranched cycloalkanes, and information on extension to graphs containing multiple edges and/or loops (molecules containing multiple bonds and/or heteroatoms), if available. The first author who explicitly used the term molecular complexity index was Bertz [30] in 1981.

2.3.1 The Minoli Index

Minoli [65] proposed in 1975 an index to measure the combinatorial complexity of graphs. The Minoli index, denoted by MI, is defined as:

$$\mathrm{MI} = [\mathrm{V} \cdot \mathrm{E}/(\mathrm{V} + \mathrm{E})] \sum_{\ell} \mathrm{P}_{\ell} \tag{2.1}$$

where V and E are the number of vertices and edges, respectively, and P_{ℓ} is the number of paths of length ℓ (a path is a sequence of adjacent edges which do not pass through the same vertex more than once). Note that $\mathrm{P}_1 = \mathrm{E}$. Computation of the Minoli index is illustrated in Figure 2.4.

The Minoli index can be given in closed form for the alkanes:

$$\mathrm{MI} = [\mathrm{V}^2 (\mathrm{V} - 1)^2] / (4\mathrm{V} - 2) \tag{2.2}$$

and cycloalkanes:

$$\mathrm{MI} = [\mathrm{V}^2 (\mathrm{V} - 1)] / 2 \tag{2.3}$$

The Minoli index is a highly degenerate index: it does not discriminate between isomeric acyclic structures. In order to increase the discriminating ability of the index, Bonchev [42] in 1990 replaced the sum of the number of paths by the total length of all paths Λ_{p}. The total length of all paths in G is:

$$\Lambda_{\mathrm{p}} = \sum_{\ell} \ell \cdot \mathrm{P}_{\ell} \tag{2.4}$$

The Bonchev variant of the Minoli index, which we shall call the Minoli–Bonchev index, denoted by MB, is defined as:

$$\mathrm{MB} = [\mathrm{V} \cdot \mathrm{E}/(\mathrm{V} + \mathrm{E})]\, \Lambda_{\mathrm{p}} \tag{2.5}$$

An exemplary calculation of the Minoli–Bonchev index is given in Figure 2.5.

Closed formulas can be given for the Minoli–Bonchev index for *n*-alkanes:

$$\mathrm{MB} = (\mathrm{V} - 1)^2\, \mathrm{V}^2\, (\mathrm{V} + 1)/6(2\mathrm{V} - 1) \tag{2.6}$$

and for cycloalkanes:

$$\mathrm{MB} = (\mathrm{V} - 1)\, \mathrm{V}^3/4 \tag{2.7}$$

(a) Graph G representing methylcyclopropane

$V = 4$

$E = 4$

G

(b) The path counts in G

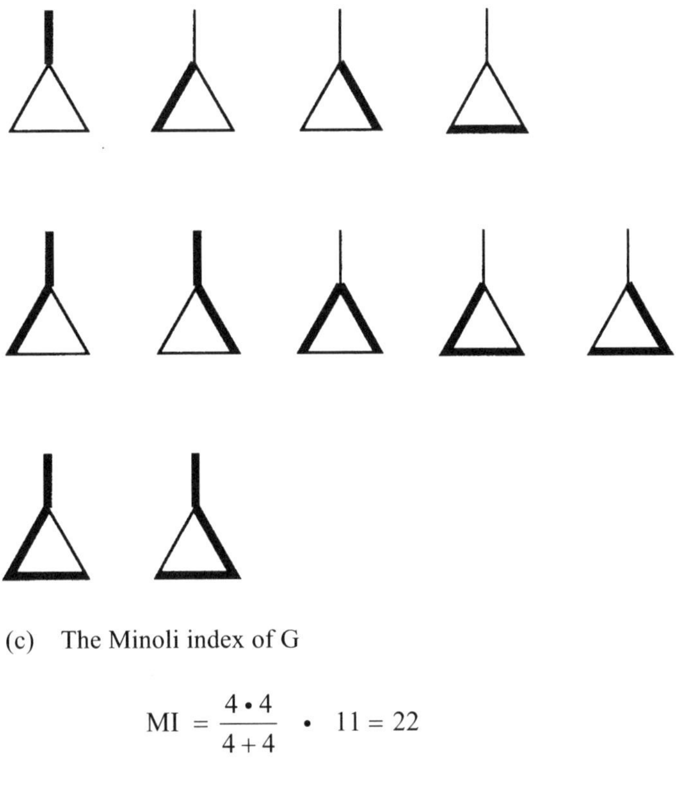

$P_1 = 4$

$P_2 = 5$

$P_3 = 2$

(c) The Minoli index of G

$$MI = \frac{4 \bullet 4}{4 + 4} \bullet 11 = 22$$

Figure 2.4: Computation of the Minoli index (MI) for the hydrogen-suppressed methylcyclopropane graph.

(a) Methylcyclopropane graph

G

(b) The total length of all paths in G

$$\Lambda_p = 1 \bullet P_1 + 2 \bullet P_2 + 3 \bullet P_3 = 20$$

(c) The Minoli–Bonchev index of G

$$MB = \frac{4 \bullet 4}{4+4} \bullet 20 = 40$$

Figure 2.5: Computation of the Minoli–Bonchev index (MB) for the hydrogen-suppressed methylcyclopropane graph.

As already stated, the Minoli index is not sufficiently structure-sensitive since the total path count for isomeric alkanes is the same, e.g., for *n*-pentane and 2,2-dimethylpropane $\sum P_\ell = 10$. On the other hand, the Minoli–Bonchev index is more structure-sensitive in that the value is smaller for more branched structures, e.g., $\Lambda_p = 20$ for *n*-pentane and $\Lambda_p = 16$ for 2,2-dimethylpropane. As seen in this example, MB decreases with increasing branching. This follows from its positive association with path lengths.

To calculate a Minoli or Minoli–Bonchev index value requires knowledge of all paths in a graph (molecule). Often it is not easy to find all paths manually, and so path finding computer programs may be helpful [73].

2.3.2 Information-Theoretic Indices

Information-theoretic indices are derived from the Shannon formula for the information content of a system having n elements [66,67]:

$$I = -\sum_{i=1}^{k} (n_i / n) \log_2 (n_i / n) \tag{2.8}$$

where k is the number of different sets of elements, n_i is the number of elements in the i-th set and summation is over all sets of elements. The ratio n_i/n (usually denoted by p_i) is the probability for a randomly selected element of the system to be found in the i-th set of elements. All Shannon-derived indices have to partition some features of the graph into equivalence classes (sets). They differ as to which features are partitioned (vertices or distances or pairs of adjacent edges, etc.) and as to the criteria of partition (chemical nature of vertex, degree of vertex, degree of vertex and its first neighbors, equivalence of vertices by symmetry, length of distance, equivalence of vertex pairs by symmetry, equivalence of pairs of adjacent edges by symmetry, etc.). It should also be noted that Shannon-derived indices are symmetry-sensitive by definition, since the Shannon formula measures the degree of uniformity, i.e., symmetry largely governs the partition.

Rashevsky was the first to calculate the information content of graphs [68(a)] and to apply this concept to the graph of metabolic reactions of a living organism [68(b)]. While he used for the partition of vertices such obvious criteria as the degree and the distance of a vertex from an easily distinguished landmark vertex, Trucco [69] and Mowshowitz [8] for the same purpose used the complete graph-theoretical information as expressed in the orbits of the graph's automorphism group. More recently, Basak [39] derived several topological indices using the Shannon formula and employed them in structure-property-activity modeling [70]. Here we will review the information-theoretic indices due to Bonchev and Trinajstić [71] and Bertz [1,30–34] that were used to quantify the complexity of molecules.

2.3.2.1 The Bonchev–Trinajstić Index Bonchev and Trinajstić [71] derived in 1977 two indices, using information-theoretic tools [66,72], to study the concept of molecular branching. One of these indices, denoted as the BT index, is obtained from equation (2.8) by multiplication by n and choosing as objects to be partitioned the distances:

$$\mathrm{BT} = \mathrm{n} \log_2 \mathrm{n} - \sum_{\ell} \mathrm{n}_\ell \log_2 \mathrm{n}_\ell \tag{2.9}$$

where n is the total number of distances, n_ℓ the count of distances of length ℓ, and $\mathrm{n} = \sum \mathrm{n}_\ell$. The calculation of a BT index value is exemplified in Figure 2.6.

Generally, $\mathrm{n} = \binom{V}{2}$, and for *n*-alkanes a formula can be given:

$$\mathrm{BT} = \binom{V}{2} \log_2 \binom{V}{2} - \sum_{i=2}^{V-1} i \log_2 i \tag{2.10}$$

(a) Graph G representing 1,2-dimethylcyclopentane

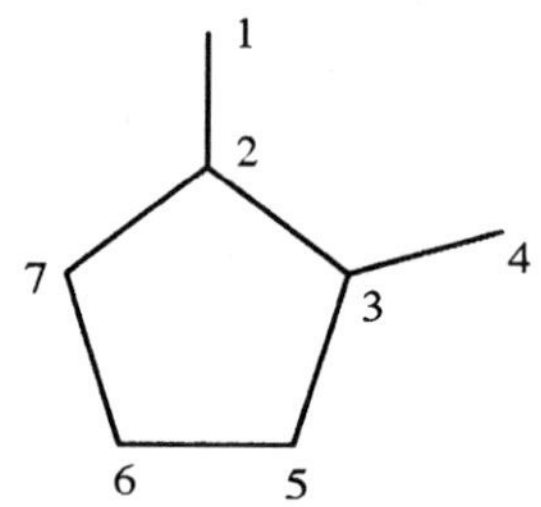

G

(b) Enumeration of distances in G using the upper triangle of the corresponding distance matrix

$$\begin{bmatrix} 0 & 1 & 2 & 3 & 3 & 3 & 2 \\ & 0 & 1 & 2 & 2 & 2 & 1 \\ & & 0 & 1 & 1 & 2 & 2 \\ & & & 0 & 2 & 3 & 3 \\ & & & & 0 & 1 & 2 \\ & & & & & 0 & 1 \\ & & & & & & 0 \end{bmatrix} \qquad \begin{aligned} n &= 21 \\ n_1 &= 7 \\ n_2 &= 9 \\ n_3 &= 5 \end{aligned}$$

(c) The BT index of G

$$BT = n \log_2 n - (n_1 \log_2 n_1 + n_2 \log_2 n_2 + n_3 \log_2 n_3) = 32.45$$

Figure 2.6: Computation of the Bonchev–Trinajstić index (BT) for the hydrogen-suppressed 1,2-dimethylcyclopentane graph.

For unbranched cycloalkanes:

$$\mathrm{BT} = \begin{cases} \binom{V}{2} \log_2 (V-1) - V(V-2)/2 & \text{if V is even} \\ \binom{V}{2} \log_2 [(V-1)/2] & \text{if V is odd} \end{cases} \tag{2.11}$$

The BT values for *n*-pentane and 2,2-dimethylpropane are, respectively, 18.47 and 9.71. The BT index for the most symmetric graphs, that is, for complete graphs, is always zero, since in complete graphs all distances are identical and equal to unity.

2.3.2.2 The Bertz Index The Bertz index, denoted by BI, is defined as [1,30–34]:

$$\mathrm{BI} = 2\,n \log_2 n - \sum_i n_i \log_2 n_i \tag{2.12}$$

where n stands for a selected graph invariant. Bertz found that a good invariant to use is the number of pairs of adjacent edges in a molecular graph G. Thus, n is the total number of pairs of adjacent edges (formerly denoted η) and n_i is the number of pairs of adjacent edges in the i-th class by symmetry. The number n can be obtained, for example, by counting the edges in the line graph [22,24] (called by Bertz the bond graph) generated from G [31]. Number n in a molecular graph G is the number of paths of length 2. A very convenient calculation of n uses the following formula [33,34,40]:

$$n = \sum_{i=1}^{V} \binom{d_i}{2} \tag{2.13}$$

where d_i is the degree of vertex i.

To use BI requires knowledge of the numbers n_i, that is, of the graph's symmetry. Numbers n_i can be obtained, for example, by the symmetry program TOPSYM [74].

Equation (2.12) represents a pragmatic modification of equation (2.9). Bertz added to equation (2.9) the extra term $n \log_2 n$ to prevent the result BI = 0 when all pairs of adjacent edges in a graph are equivalent, as is the case for graphs depicting propane, isobutane, cycloalkanes, benzene, the annulenes, etc. This modification of equation (2.9) explicitly takes into account the size of the molecule. The first part of equation (2.12) roughly measures structural characteristics of G (size, branching, cyclicity, etc.), while the second part reflects the symmetry of G in terms of equivalent pairs of adjacent edges. Therefore, the Bertz index represents a balance between structural features and the symmetry of a molecular graph. Bertz introduced his approach for use in synthesis planning and analysis of synthetic strategies [1,30,32,43]. Computation of the Bertz index is exemplified in Figure 2.7.

For *n*-alkanes the general formula can be given:

$$\mathrm{BI} = \begin{cases} 2(\mathrm{V}-2)\log_2(\mathrm{V}-2) - (\mathrm{V}-2) & \text{if V is even} \\ 2(\mathrm{V}-2)\log_2(\mathrm{V}-2) - (\mathrm{V}-3) & \text{if V is odd} \end{cases} \tag{2.14}$$

For unbranched cycloalkanes:

$$\mathrm{BI} = \mathrm{V}\log_2 \mathrm{V} \tag{2.15}$$

The Bertz index was applied to molecules containing multiple bonds [31,40] and heteroatoms [30,40,41].

2.3.3 The Number of Spanning Trees

The first authors who used spanning trees ("maximal trees") explicitly as a measure of complexity were Bonchev, Temkin and Kamenski [29] in 1980 and in more detail [38] in 1987. Bonchev and coworkers also used the number of spanning trees t(G) to study the complexity of cyclic graphs [45,51]. Independently, Mallion and coworkers used the number of spanning trees as a criterion for the complexity of (poly)cyclic graphs [75, 76]: The most complex structure possesses the highest number of spanning trees. A spanning tree of a graph G is a connected acyclic subgraph containing all vertices of G. In the case of trees, the spanning tree is identical to the tree itself. Therefore, the number of spanning trees can be used as a measure of complexity for cyclic structures only. Trivially,

$$\mathrm{t}(n\text{-alkane}) = 1 \quad \text{and} \quad \mathrm{t}(\text{cycloalkane}) = \mathrm{V} \tag{2.16}$$

A lot of work on the enumeration of spanning trees for various classes of (molecular) graphs has been carried out over the last 25 years [76 and references cited therein]. The interest in spanning trees is related to several chemical problems such as those involving chemical kinetics [29, 38, 45, 51] and calculation of the magnetic properties of conjugated systems by means of the ring-current model within the framework of π-electron molecular orbital theory [76, 77 and references cited therein].

(a) 1,2-dimethylcyclopentane graph

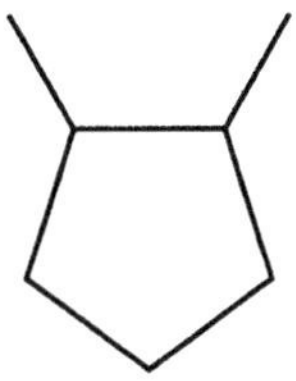

G

(b) Counting pairs of adjacent edges in G

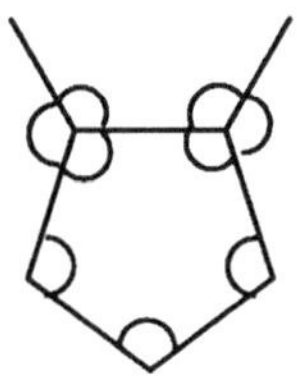

n = 9

(c) Classes of pairs of adjacent edges as determined by symmetry

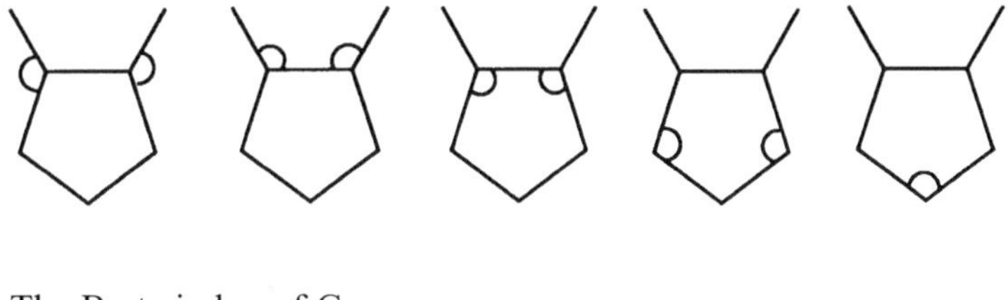

(d) The Bertz index of G

$$\text{BI} = 2 \bullet 9 \log_2 9 - \left(1 \log_2 1 + 4 \bullet 2 \log_2 2\right) = 49.06$$

Figure 2.7: Computation of the Bertz index (BI) for the graph G representing 1,2-dimethylcyclopentane.

The number of spanning trees can be obtained in several ways [52]. Here we give four approaches based mainly on the Laplacian matrix. The Laplacian matrix **L** = **L** (G) of a graph G is defined as [77 and references cited therein]:

$$\mathbf{L} = \mathbf{V} - \mathbf{A} \tag{2.17}$$

where **V** is a diagonal matrix whose entries are the vertex degrees and **A** is the adjacency matrix of G. The Laplacian matrix is a real symmetric matrix which has found use, for example, in structure-property modeling through topological indices derived from it [77]. The diagonalization of the Laplacian matrix of G with V vertices produces V real eigenvalues. The set of Laplacian eigenvalues is usually referred to as the Laplacian spectrum and is denoted by $x_i(\mathbf{L})$. The smallest member of the Laplacian spectrum is always zero. This is a consequence of the special structure of the Laplacian matrix.

2.3.3.1 Computation based on the determinant of a Laplacian submatrix
The most important application of the Laplacian matrix of a graph G is in the matrix-tree theorem for spanning trees [78]:

$$t(G) = |\,\mathbf{L}_{ij}\,| \tag{2.18}$$

where t(G) is the number of spanning trees of G and $\mathbf{L}_{ij}$ the submatrix of the Laplacian matrix **L** derived by deleting row i and column j. The use of equation (2.18) is illustrated in Figure 2.8. The corresponding spanning trees are depicted in Figure 2.9.

2.3.3.2 Computation based on the spectrum of the Laplacian matrix
The number of spanning trees t(G) in terms of the Laplacian matrix spectrum $x_i(\mathbf{L})$ is given by [77]:

$$t(G) = (1/V)\prod_{i=2}^{V} x_i(\mathbf{L}) \tag{2.19}$$

The use of this procedure for computing the number of spanning trees is exemplified in Figure 2.10.

2.3.3.3 Computation based on the characteristic polynomial of an inner dual of a planar graph This method consists of the following steps [75]: Construction of an inner dual G′ of G (an inner dual G′ of G is obtained by placing a vertex in each ring of a planar polycyclic graph; a pair of vertices is connected if the corresponding rings in G have an edge in

(a) Graph G representing benzocyclobutadiene

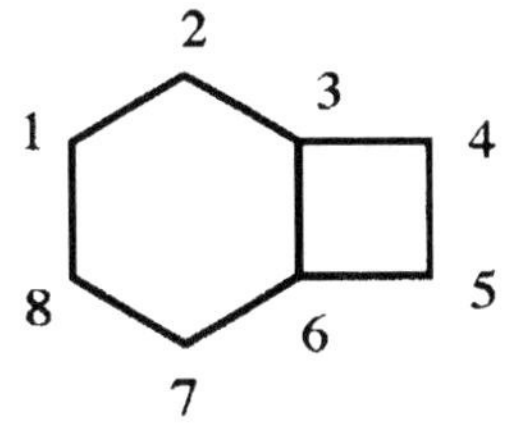

G

(b) Laplacian matrix of G

$$\mathbf{L} = \begin{bmatrix} 2 & -1 & 0 & 0 & 0 & 0 & 0 & -1 \\ -1 & 2 & -1 & 0 & 0 & 0 & 0 & 0 \\ 0 & -1 & 3 & -1 & 0 & -1 & 0 & 0 \\ 0 & 0 & -1 & 2 & -1 & 0 & 0 & 0 \\ 0 & 0 & 0 & -1 & 2 & -1 & 0 & 0 \\ 0 & 0 & -1 & 0 & -1 & 3 & -1 & 0 \\ 0 & 0 & 0 & 0 & 0 & -1 & 2 & -1 \\ -1 & 0 & 0 & 0 & 0 & 0 & -1 & 2 \end{bmatrix}$$

(the first row, $i=1$, and the first column, $j=1$, are struck out)

(c) Submatrix $\mathbf{L}_{11}$ of the Laplacian matrix $\mathbf{L}$ of G

$$\mathbf{L}_{11} = \begin{bmatrix} 2 & -1 & 0 & 0 & 0 & 0 & 0 \\ -1 & 3 & -1 & 0 & -1 & 0 & 0 \\ 0 & -1 & 2 & -1 & 0 & 0 & 0 \\ 0 & 0 & -1 & 2 & -1 & 0 & 0 \\ 0 & -1 & 0 & -1 & 3 & -1 & 0 \\ 0 & 0 & 0 & 0 & -1 & 2 & -1 \\ 0 & 0 & 0 & 0 & 0 & -1 & 2 \end{bmatrix}$$

(d) The number of spanning trees of G

$$t(G) = |\,\mathbf{L}_{11}| = 23$$

Figure 2.8: The use of the matrix-tree theorem to compute the number of spanning trees for the hydrogen-suppressed benzocyclobutadiene graph.

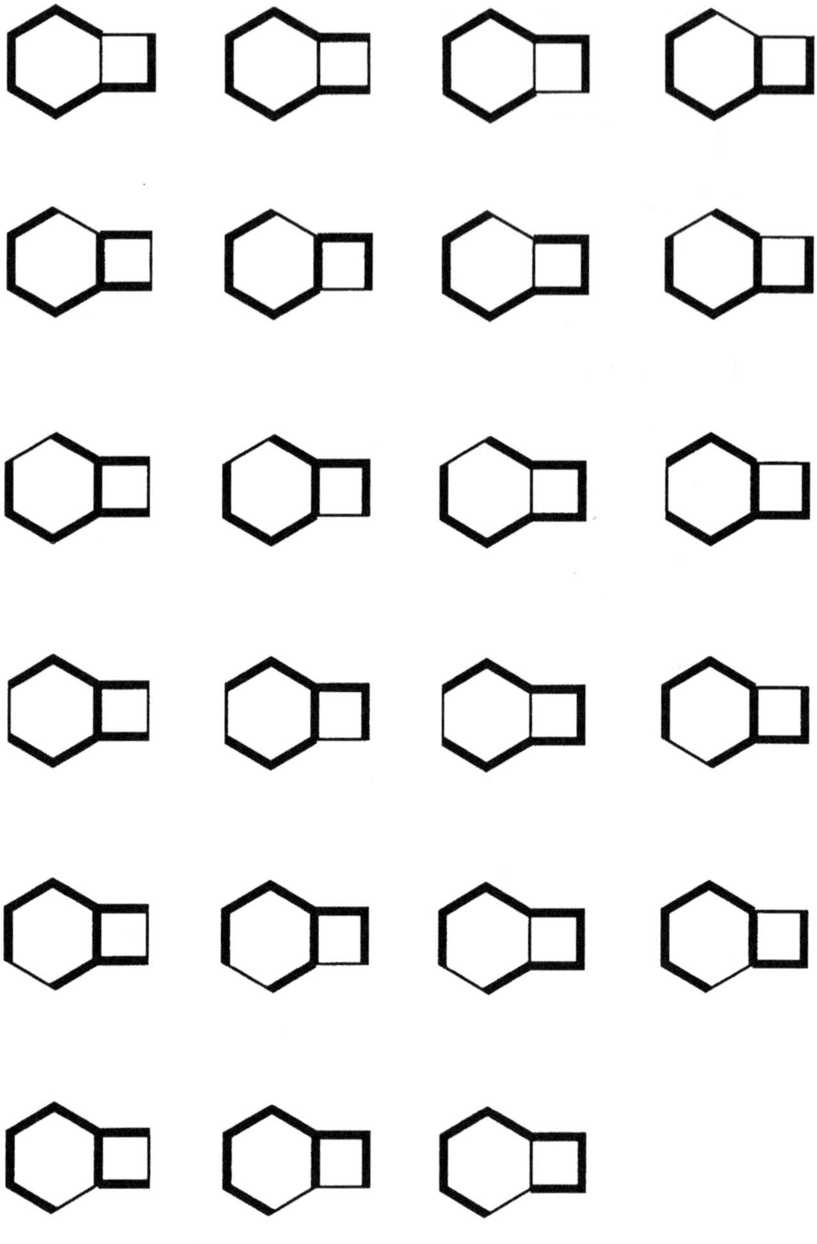

Figure 2.9: All spanning trees belonging to the benzocyclobutadiene graph.

(a) Benzocyclobutadiene graph

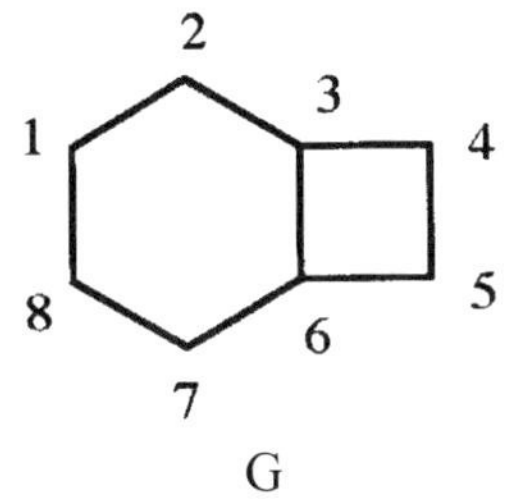

(b) Laplacian matrix of G

$$\mathbf{L} = \begin{bmatrix} 2 & -1 & 0 & 0 & 0 & 0 & 0 & -1 \\ -1 & 2 & -1 & 0 & 0 & 0 & 0 & 0 \\ 0 & -1 & 3 & -1 & 0 & -1 & 0 & 0 \\ 0 & 0 & -1 & 2 & -1 & 0 & 0 & 0 \\ 0 & 0 & 0 & -1 & 2 & -1 & 0 & 0 \\ 0 & 0 & -1 & 0 & -1 & 3 & -1 & 0 \\ 0 & 0 & 0 & 0 & 0 & -1 & 2 & -1 \\ -1 & 0 & 0 & 0 & 0 & 0 & -1 & 2 \end{bmatrix}$$

(c) Laplacian spectrum of G

{0.0000, 0.5858, 1.0681, 2.0000, 2.4824, 3.4142, 3.5176, 4.9319}

(d) The number of spanning trees of G

$$t(G) = \frac{1}{8}\prod_{i=2}^{8} x_i\ (\mathbf{L}) = 23$$

Figure 2.10: Computation of the number of spanning trees for the benzocyclobutadiene graph G using equation (2.19).

common), computation of the characteristic polynomial of G′, P(G′; x, y, …) and substitution of the ring sizes into the characteristic polynomial of G′, P(G′, x, y, … = ring sizes). Schematically, this can be represented as G → G′ → P(G′, x,y,..) → P(G′, x,y, … = ring sizes) = t(G). This procedure is illustrated in Figure 2.11.

(a) Benzocyclobutadiene graph

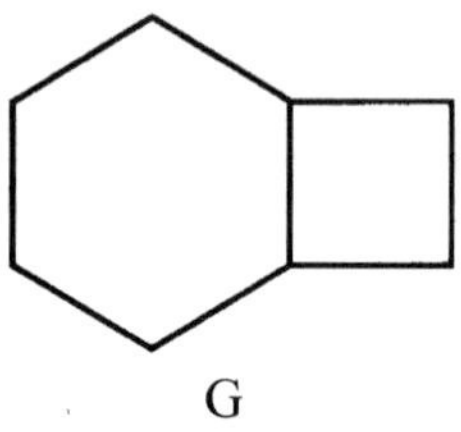

(b) Inner dual G′ of G

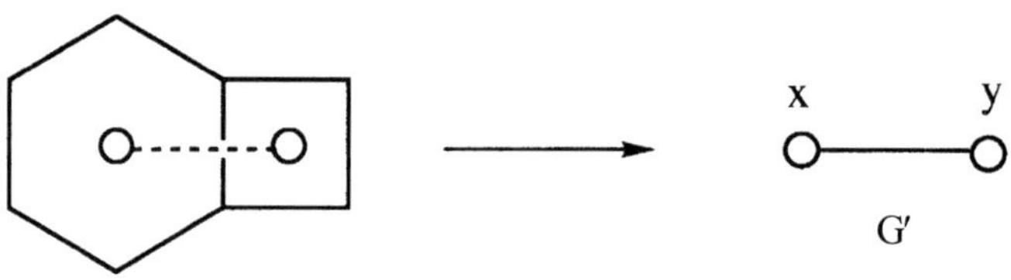

(c) Characteristic polynomial of G′

$$P(G'; x, y) = xy - 1$$

(d) The number of spanning trees of G

$$t(G) = P(G'; x = 6, y = 4) = 23$$

Figure 2.11: Computation of the number of spanning trees for the benzocyclobutadiene graph G using the inner dual G′ of G. The inner dual G′ is a graph whose vertices are weighted with the sizes of the corresponding rings in G.

2.3.3.4 Computation based on the coefficient at the x term of the Laplacian polynomial of G The relationship between the number of spanning trees and the absolute value of the coefficient at the x term of the Laplacian polynomial of G is given by [79]:

$$t(G) = |\text{Coefficient of } x| / V \tag{2.20}$$

The origin of this simple formula is the fact that the coefficient at the x term of the Laplacian polynomial must be equal to the product of all non-zero Laplacian eigenvalues:

(a) Benzocyclobutadiene graph

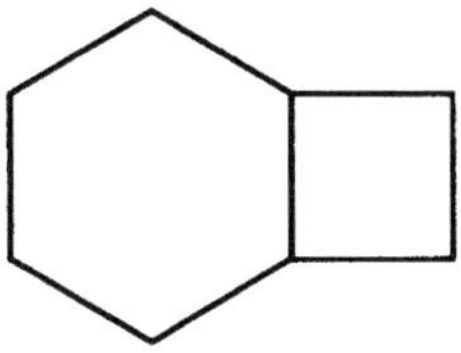

G

(b) Laplacian polynomial of G

$$P(G; x) = x^8 - 18x^7 + 132x^6 - 508x^5 + 1098x^4 - 1316x^3 + 796x^2 - 184x$$

(c) The number of spanning trees of G

$$t(G) = 184 / 8 = 23$$

Figure 2.12: Computation of the number of spanning trees for the benzocyclobutadiene graph using equation (2.20).

$$\text{Coefficient of } x = \prod_{i=2}^{V} x_i(\mathbf{L}) \tag{2.21}$$

This approach is useful if the Laplacian polynomials are available. The use of equation (2.21) is demonstrated in Figure 2.12.

The fastest way to compute the number of spanning trees of any graph is via equation (2.18), and software packages such as Mathematica or Maple can be used for computing t(G) efficiently.

It should be noted that the number of spanning trees is a rather explosive combinatorial quantity even for moderately large graphs. For example, the modestly sized kekulene graph has 16712665200 spanning trees [52].

2.3.4 The Number of Kinds of Connected Subgraphs and the Total Number of Connected Subgraphs

Early work on counting and using subgraphs in chemistry was carried out by Gordon and Kennedy [80]. They introduced a graph invariant N_{sG} which is defined as the number of distinct ways in which a subgraph s can be cut out of a graph G. Bertz and Herndon in 1986 first explored the same idea for measuring molecular similarity and complexity [36]. A connected subgraph is obtained by removing any one or more edges and/or vertices from a graph such that the remainder is still one piece (the graph itself is counted as a subgraph for formal reasons [21, 22]). Obviously, in this way, many isomorphic connected subgraphs are produced and therefore not only the total number of connected subgraphs N_t, but also the number of non-isomorphic connected subgraphs N_s ("the number of kinds of connected subgraphs") are interesting graph invariants. Though not very sensitive to branching, N_s is valuable to chemists since by its very nature it is sensitive to multiple bonds and heteroatoms [54] as well as to symmetry, and is independent of the particular numerical expression chosen for describing these features. Bertz and coworkers used N_t and N_s for the study of synthetic strategies [1,54], while Bonchev independently used N_t for expressing molecular complexity and for structure-property modeling [56, 63(b)].

In Figure 2.13 we exemplify the generation and counting of subgraphs to obtain N_s and N_t for the methylcyclopropane graph.

Trivially, the number of methane subgraphs of G equals the number of vertices, $N_m = V$, and the number of ethane subgraphs equals the number of edges, $N_e = E$. The following closed formulas are easily derived for *n*-alkanes:

$$N_t = V(V + 1)/2 \tag{2.22}$$

$$N_s = V \tag{2.23}$$

and for unbranched cycloalkanes:

$$N_t = V^2 + 1 \tag{2.24}$$

$$N_s = V + 1 \tag{2.25}$$

While the above researchers constructed the subgraphs manually, Bone and Villar attempted to enumerate all substructures of a molecular structure by computer [81]. Unfortunately, they defined a connected subgraph simply as a set of connected vertices, thus necessarily missing all those subgraphs in which the same set of vertices is connected by different sets of edges. Therefore, in the published lists of substructures found for their worked examples (methylcyclopentane and 3-methyltetrahydrofuran) a few substructures that should appear are missing.

Two of us (G.R., C.R.) therefore decided to write a computer program for the generation of all connected subgraphs of a given connected undirected

(a) Graph G representing methylcyclopropane

G

(b) Enumeration of all connected subgraphs of G

(b.1) The number of methane subgraphs

$N_m = 4$

(b.2) The number of ethane subgraphs

$N_e = 4$

(b.3) The number of propane subgraphs

$N_p = 5$

(b.4) The number of butane subgraphs

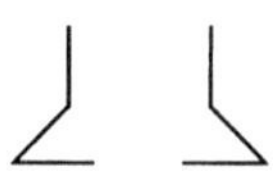

$N_b = 2$

(b.5) The number of isobutane subgraphs

$N_i = 1$

(b.6) The number of cyclopropane subgraphs

$N_c = 1$

(b.7) Graph G as its own subgraph

$N_G = 1$

(c) The N_S and N_t numbers of G

$N_S = 7$ $N_t = 18$

Figure 2.13: Computation of N_s and N_t for the methylcyclopropane graph.

simple graph, which is now available [82]. (A simple graph is a graph without multiple edges or loops.) A connected subgraph can be described as a string of labels of adjacent edges, just as a path is a string of labels of adjacent vertices. The connected subgraphs therefore are constructed by a path-finding algorithm working on the graph's edge adjacency matrix, just as paths are constructed by such an algorithm working on the usual vertex adjacency matrix. As is well-known for path algorithms, the effort required (CPU time) increases exponentially with increasing graph size. An improved program now finds all substructures and all subgraphs of compounds containing heteroatoms and multiple bonds [90–92].

Bertz and Zamfirescu [61] tested the number of special subgraphs, cliques and bicliques, instead of N_t, or the intersection numbers, i.e., the minimal number of cliques or bicliques required to cover all edges of a graph or its first or second line graph, or several other related graph invariants. This approach could be extended to graphs with multiple edges.

2.3.5 The Bonchev Indices

In the same paper [56] in which he proposed to use the total number of connected subgraphs as a molecular complexity measure, Bonchev also proposed two new complexity indices based on the connected subgraphs. The first Bonchev index, called the topological complexity index and denoted by TC, is equal to the sum of vertex degrees in all connected subgraphs of a graph:

$$TC = \sum_{s}\sum_{i} d_i(s) \tag{2.26}$$

where the first sum is over all connected subgraphs in a given molecular graph G, the second sum is over all vertices in a subgraph and d_i is the vertex degree. It should be noted that in TC the vertex degrees are taken as they are in G. If the vertex degrees are taken as they are in the respective isolated subgraph, then the corresponding index is the second Bonchev complexity index, denoted by TC1. By virtue of the definitions, the inequality TC > TC1 always holds, except for the methane graph (TC = TC1 = 0). Computation of TC and TC1 is shown in Figure 2.14 and Figure 2.15, respectively.

Closed formulas were derived [63(b)] for the *n*-alkanes:

$$TC = V\,(V-1)\,(V+4)/3 \tag{2.27}$$

$$TC1 = V\,(V-1)\,(V+1)/3 \tag{2.28}$$

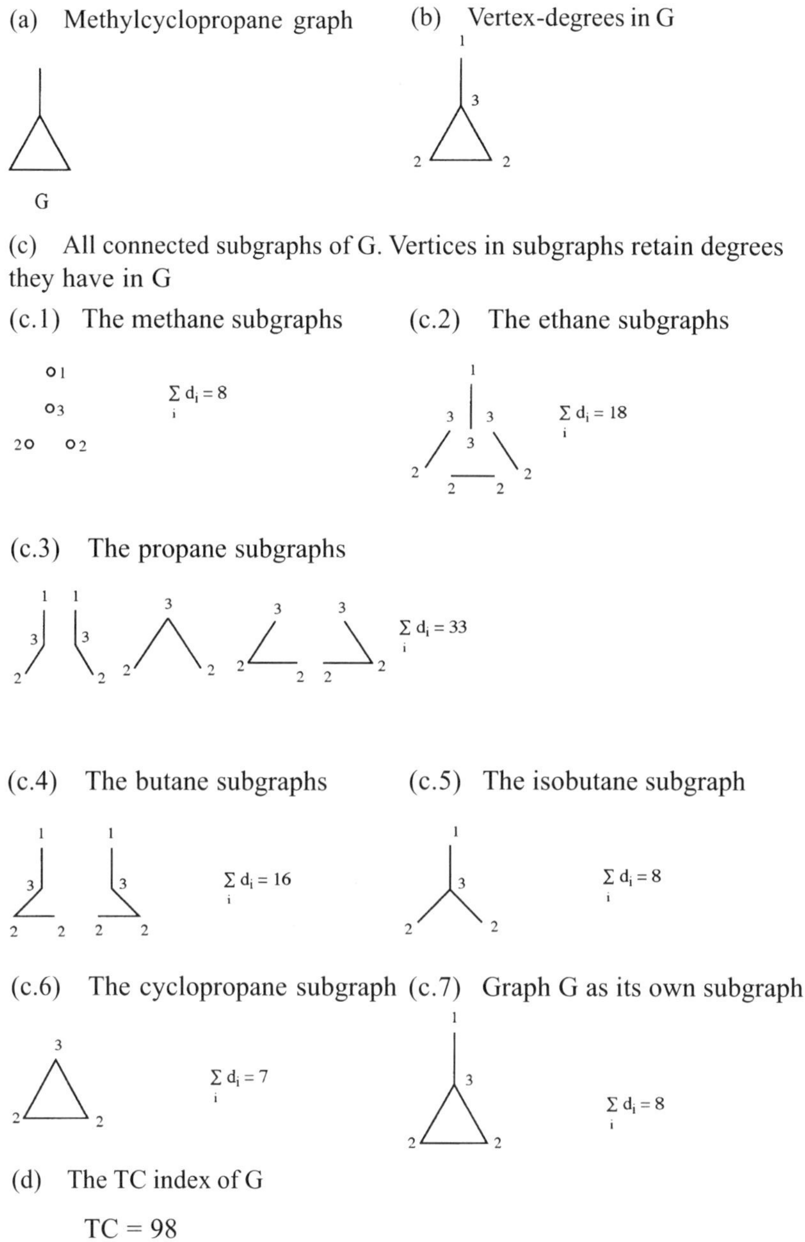

Figure 2.14: Computation of the Bonchev complexity index TC for the methylcyclopropane graph.

(a) Methylcyclopropane graph

G

(b) All connected subgraphs of G. Vertex-degrees are taken as they are in the isolated subgraphs

(b.1) The methane subgraphs

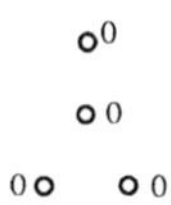

$\sum_i d_i = 0$

(b.2) The ethane subgraphs

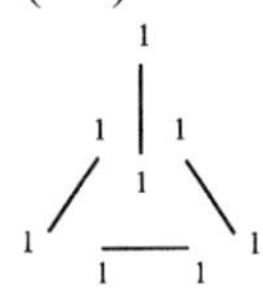

$\sum_i d_i = 8$

(b.3) The propane subgraphs

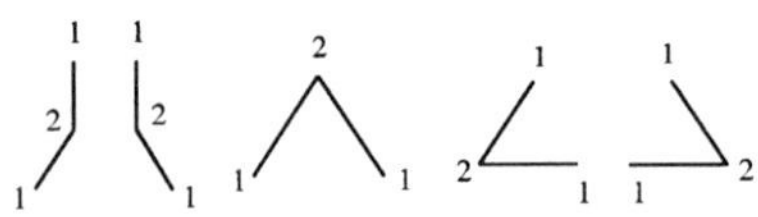

$\sum_i d_i = 20$

(b.4) The butane subgraph

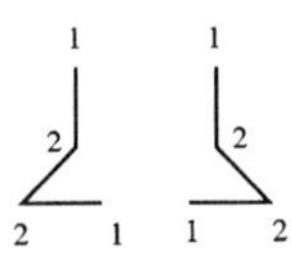

$\sum_i d_i = 12$

(b.5) The isobutane subgraphs

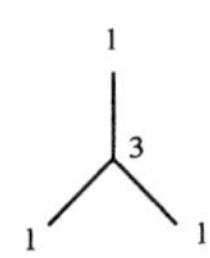

$\sum_i d_i = 6$

(b.6) The cyclopropane subgraph

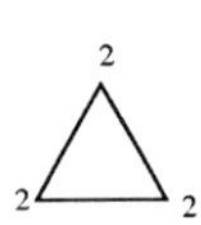

$\sum_i d_i = 6$

(b.7) Graph G as its own subgraph

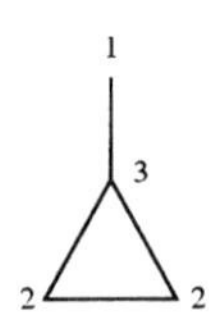

$\sum_i d_i = 8$

(c) The TC1 index of G

TC1 = 60

Figure 2.15: Computation of the Bonchev complexity index TC1 for the methylcyclopropane graph.

and cycloalkanes:

$$TC = V\ (V^2 + V + 2) \tag{2.29}$$

$$TC1 = V\ (V^2 - V + 2) \tag{2.30}$$

TC and TC1 were also defined for molecules containing multiple bonds and/or heteroatoms [63(b)].

2.3.6 The Randić Complexity Index

The Randić complexity index [59], denoted by RCI, is based on the concept of the augmented vertex degree [60]. The augmented degree for a given vertex is obtained by adding to its degree a half of the degrees of the nearest neighbors, a quarter of the degrees of the next nearest neighbors, and so on. The complexity index RCI is then given as the sum of the augmented degrees of vertices not equivalent by symmetry.

The Randić complexity index can be computed by means of the augmented degree matrix **D′**:

$$(\mathbf{D}')_{ij} = d_j / 2^{\ell_{ij}} \tag{2.31}$$

where d_j is the degree of vertex j, and ℓ_{ij} the distance between vertices i and j. The augmented degree matrix is a non-symmetric matrix. The augmented degree of vertex i is the row sum of row i of this matrix. The approach of Randić is illustrated in Figure 2.16.

Closed formulas can be derived for RCI of the *n*-alkanes:

$$RCI = \begin{cases} 3V - 6 + 6/2^V & \text{if V is even} \\ 3V - 3 - 3(2^{(V-1)/2} - 1)/2^{V-1} & \text{if V is odd} \end{cases} \tag{2.32}$$

and cycloalkanes:

$$RCI = \begin{cases} 6(2^{V/2} - 1)/2^{V/2} & \text{if V is even} \\ 6 - 1/2^{(V-5)/2} & \text{if V is odd} \end{cases} \tag{2.33}$$

Thus, RCI(*n*-alkane) for increasing V converges toward 3V – 6 for even V and toward 3V – 3 for odd V, and RCI(cycloalkane) converges toward 6.

(a) Bicyclo[2.1.0]pentane graph, labeled arbitrarily

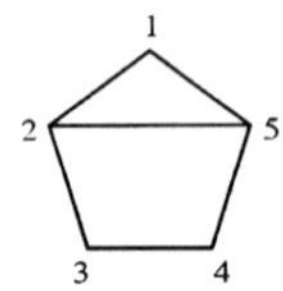

G

(b) Augmented degree matrix of G

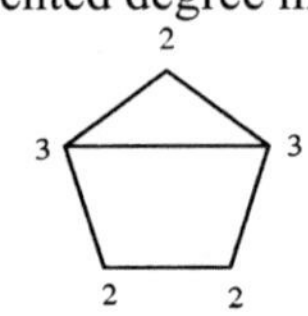

Vertex degrees

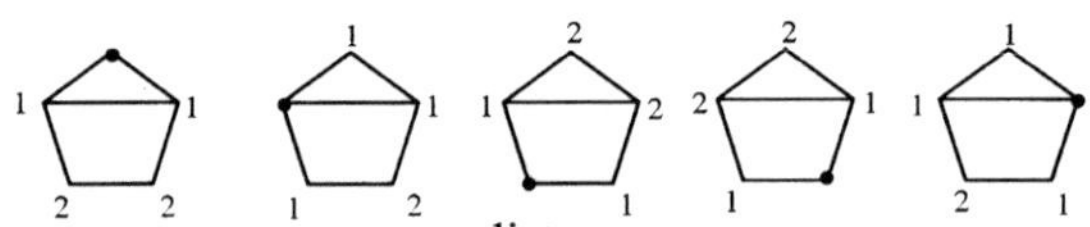

distances

$$\mathbf{D}' = \begin{bmatrix} 2 & 3/2 & 2/4 & 2/4 & 3/2 \\ 2/2 & 3 & 2/2 & 2/4 & 3/2 \\ 2/4 & 3/2 & 2 & 2/2 & 3/4 \\ 2/4 & 3/4 & 2/2 & 2 & 3/2 \\ 2/2 & 3/2 & 2/4 & 2/2 & 3 \end{bmatrix}$$

(c) Augmented vertex degrees (AV) for symmetry non-equivalent vertices of G

$$AV(1) = 2 + 6/2 + 4/4 = 6$$
$$AV(2) = 3 + 7/2 + 2/4 = 7$$
$$AV(3) = 2 + 5/2 + 5/4 = 5.75$$

(d) The Randić complexity index of G

$$\begin{aligned} RCI &= AV(1) + AV(2) + AV(3) \\ &= 6 + 7 + 5.75 \\ &= 18.75 \end{aligned}$$

Figure 2.16: Computation of the Randić complexity index (RCI) for the bicyclo[2.1.0] pentane graph.

2.3.7 The Zagreb Complexity Indices

We will not be concerned here with the standard topological indices employed in structure-property-activity modeling [63], except for the so-called Zagreb indices [1, 24, 62]:

$$M_1 = \sum_{\text{vertices}} d_i^2 \tag{2.34}$$

$$M_2 = \sum_{\text{edges}} d_i\, d_j \tag{2.35}$$

where d_i is the degree of vertex i. Bertz and Wright [1] pointed out that the Zagreb M_1 index has the characteristics of a complexity index.

M_1 equals the count of walks (see below) of length 2 (mwc_2) in a graph:

$$M_1 = mwc_2 \tag{2.36}$$

and is related to the number of self-returning walks of length 4 (^4SRW) [48]:

$$M_1 = (^4SRW + 2\,E) / 2 \tag{2.37}$$

While the former equation is generally valid, the latter is limited to graphs without four-membered rings.

The M_1 and M_2 indices can be given in closed form for homologous structures. Here we give analytical formulas for several classes of regular structures:

(i) The *n*-alkanes:

$$M_1 = 4\,(V - 2) + 2 \tag{2.38}$$

$$M_2 = 4\,(V - 2) \text{ for } V > 2 \tag{2.39}$$

(ii) The cycloalkanes:

$$M_1 = M_2 = 4V \tag{2.40}$$

(iii) The polyacenes:

$$M_1 = 26\,R - 2 \tag{2.41}$$

$$M_2 = 33\,R - 9 \tag{2.42}$$

(iv) The polyphenanthrenes (zig-zag benzenoids):

$$M_1 \text{ (polyphenanthrene)} = M_1 \text{ (polyacene)} \tag{2.43}$$

$$M_2 = 34\,R - 11 \tag{2.44}$$

where R is the number of hexagons in a polyacene or polyphenanthrene. The equality $M_1 = M_2$ for cycloalkanes is a consequence of the fact that in these molecules the number and connectivity of the carbon atoms is equal to the number and connectivity of the carbon-carbon bonds. The equality M_1 (polyphenanthrene) = M_1(polyacene) is a consequence of the definition of the M_1 index. Index M_1 depends only on the valencies of atoms (degrees) in a molecule (graph), and these are the same in the isomeric benzenoids.

Computation of the M_1 and M_2 indices for the 2,3,3-trimethylhexane tree is shown in Figure 2.17.

If we combine the idea of Bonchev, used to derive the TC and TC1 indices, with the Zagreb indices, we arrive at four new complexity indices (TM_1, TM_1^*, TM_2, TM_2^*). The TM_1 (TM_2) index is equal to the sum of the M_1 (M_2) indices for all connected subgraphs of a graph:

$$TM_1 = \sum_{s} \sum_{\text{vertices}} d_i^2(s) \tag{2.45}$$

$$TM_2 = \sum_{s} \sum_{\text{edges}} d_i\, d_j(s) \tag{2.46}$$

It should be noted that in TM_1, the M_1 index is computed for each subgraph using the vertex degrees as they are in the graph G. If the M_1 index is computed for each subgraph using the vertex degrees as they are in the isolated subgraphs, then the corresponding complexity index is TM_1^*. In the case of the TM_2 index, the M_2 index is computed for each subgraph using the edge weights as they are in the graph G. If the M_2 index is computed for each subgraph using the edge weights as they are in the isolated subgraphs, then the corresponding complexity index is TM_2^*. Note that always $TM_1(TM_2) > TM_1^*(TM_2^*)$, except for $TM_1 = TM_1^* = 0$ for methane and $TM_2 = TM_2^*$ for ethane. For example, the values of the TM_1, TM_1^*, TM_2 and TM_2^* indices for the methylcyclopropane graph are, respectively, as follows: 230, 100, 145 and 80.

(a) Branched tree T representing 2,3,3-trimethylhexane

T

(b) Vertex degrees d_i of T

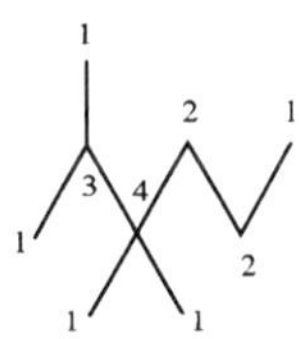

(c) The M_1 index of T

$$M_1 = \sum_{\text{vertices}} d_i^2 = 38$$

(d) Edge weights $d_i d_j$ of T

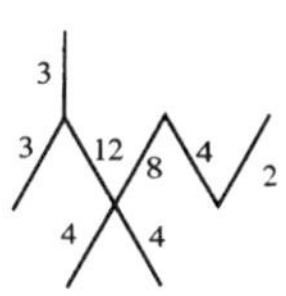

(e) The M_2 index of T

$$M_2 = \sum_{\text{edges}} d_i d_j = 40$$

Figure 2.17: Computation of the Zagreb indices M_1 and M_2 for the 2,3,3-trimethylhexane tree.

2.3.8 The Total Walk Count

The total walk count, denoted twc, was proposed by Rücker and Rücker [44,58,83] as a measure of the complexity of a graph or molecule, or more precisely as a measure of molecular labyrinthicity. A walk in a (molecular) graph is an alternating sequence of vertices and edges, beginning and ending with a vertex, in which each edge is incident with two vertices immediately preceding and following it. The length of the walk is the number of edges in it. Repetition of vertices and/or edges is allowed in a walk. It is well known in graph theory [21] that the number of walks of length ℓ beginning at vertex i and ending at vertex j is given by the element $(\mathbf{A}^{\ell})_{ij}$, the ij-element, in the ℓ-th power of the adjacency matrix $\mathbf{A}^{\ell}$.

The atomic walk count of length ℓ of atom i is denoted by $\mathrm{awc}_{\ell}(\mathrm{i})$ and is given by:

$$\mathrm{awc}_{\ell}(\mathrm{i}) = \sum_{j=1}^{V} (\mathbf{A}^{\ell})_{ij} \tag{2.47}$$

The atomic walk count sum of atom i, denoted awcs(i), is obtained by summing the $\mathrm{awc}_{\ell}(\mathrm{i})$ over all ℓ from 1 to V–1:

$$\mathrm{awcs(i)} = \sum_{\ell=1}^{V-1} \mathrm{awc}_{\ell}(\mathrm{i}) = \sum_{\ell=1}^{V-1}\sum_{j=1}^{V} (\mathbf{A}^{\ell})_{ij} \tag{2.48}$$

The total walk count twc is the sum of all atomic walk counts, that is:

$$\begin{aligned} \mathrm{twc} &= \sum_{i=1}^{V} \mathrm{awcs(i)} = \sum_{i=1}^{V}\sum_{\ell=1}^{V-1} \mathrm{awc}_{\ell} \\ &= \sum_{i=1}^{V}\sum_{\ell=1}^{V-1}\sum_{j=1}^{V} \left(\mathrm{A}^{\ell}\right)_{ij} \end{aligned} \tag{2.49}$$

Of course, the order of summations may be changed. Thus, the molecular walk count of length ℓ, denoted mcw_{ℓ}, is obtained by summing all walk counts of length ℓ:

$$\mathrm{mcw}_{\ell} = \sum_{i=1}^{V} \mathrm{awc}_{\ell}(i) = \sum_{i=1}^{V}\sum_{j=1}^{V} (\mathbf{A}^{\ell})_{ij} \tag{2.50}$$

The total walk count then is the sum of all mcw_{ℓ} for $\ell = 1$ to $V - 1$:

$$\mathrm{twc} = \sum_{\ell=1}^{V-1} \mathrm{mwc}_{\ell} = \sum_{\ell=1}^{V-1}\sum_{i=1}^{V}\sum_{j=1}^{V} (\mathbf{A}^{\ell})_{ij} \tag{2.51}$$

Index twc is sensitive to size, branching, cyclicity, edge weights (unsaturation) and vertex weights (heteroatoms) in the sense that it increases with an increase in all these complexity features. These properties of twc follow from its definition and were demonstrated empirically [58]. Moreover, except for regular graphs, twc is a highly discriminating index, i.e., there are only few cases of non-isomorphic graphs having the same twc, as demonstrated for alkanes [44] and for mono- and polycyclic graphs [83]. The calculation of a twc value according to equation (2.51) is shown in Figure 2.18 for the methylcyclobutane graph as an example.

For the same graph it is demonstrated in Figure 2.19 that twc can be obtained extremely easily, that is without matrix multiplication, by the Morgan algorithm: The so-called extended connectivity of vertex i (of the ℓ-th order) as calculated by repetitive summation of the ($\ell - 1$)-th order contributions of the neighbors of i, starting with the degrees, is $\mathrm{awc}_{\ell}(i)$. Thus, calculation of twc requires nothing but a sequence of addition steps.

For *n*-alkanes, cycloalkanes and regular graphs the following formulas can be derived, thanks to the regularity of these structures:

(i) The *n*-alkanes

An auxiliary integer sequence is required defined as:

$$s_i = \begin{cases} 2\binom{i}{i/2} & \text{if i is even} \\[2ex] 2\binom{i}{(i+1)/2} & \text{if i is odd} \end{cases} \tag{2.52}$$

(a) Graph G representing methylcyclobutane with arbitrary vertex numbering.
Vertex 5 is equivalent by symmetry to vertex 3.

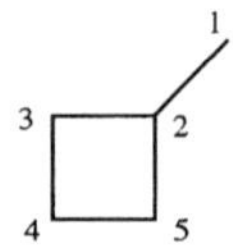

G

(b) The adjacency matrix of G

$$\mathbf{A} = \begin{bmatrix} 0 & 1 & 0 & 0 & 0 \\ 1 & 0 & 1 & 0 & 1 \\ 0 & 1 & 0 & 1 & 0 \\ 0 & 0 & 1 & 0 & 1 \\ 0 & 1 & 0 & 1 & 0 \end{bmatrix} \qquad \begin{array}{cc} 1 & 1 \\ 3 & 3 \\ 2 & 2 \\ 2 & 2 \\ 2 & \\ \hline 10 & 8 \end{array}$$

(c) The adjacency matrix powers from exponent 2 to V – 1

$$\mathbf{A}^2 = \begin{bmatrix} 1 & 0 & 1 & 0 & 1 \\ 0 & 3 & 0 & 2 & 0 \\ 1 & 0 & 2 & 0 & 2 \\ 0 & 2 & 0 & 2 & 0 \\ 1 & 0 & 2 & 0 & 2 \end{bmatrix} \qquad \begin{array}{cc} 3 & 3 \\ 5 & 5 \\ 5 & 5 \\ 4 & 4 \\ 5 & \\ \hline 22 & 17 \end{array}$$

$$\mathbf{A}^3 = \begin{bmatrix} 0 & 3 & 0 & 2 & 0 \\ 3 & 0 & 5 & 0 & 5 \\ 0 & 5 & 0 & 4 & 0 \\ 2 & 0 & 4 & 0 & 4 \\ 0 & 5 & 0 & 4 & 0 \end{bmatrix} \qquad \begin{array}{cc} 5 & 5 \\ 13 & 13 \\ 9 & 9 \\ 10 & 10 \\ 9 & \\ \hline 46 & 37 \end{array}$$

$$\mathbf{A}^4 = \begin{bmatrix} 3 & 0 & 5 & 0 & 5 \\ 0 & 13 & 0 & 10 & 0 \\ 5 & 0 & 9 & 0 & 9 \\ 0 & 10 & 0 & 8 & 0 \\ 5 & 0 & 9 & 0 & 9 \end{bmatrix} \qquad \begin{array}{cc} 13 & 13 \\ 23 & 23 \\ 23 & 23 \\ 18 & 18 \\ 23 & \\ \hline 100 & 77 \end{array}$$

(d) The total walk count for G

$$\text{twc} = 10 + 22 + 46 + 100 = 178$$
$$\text{wcx} = 8 + 17 + 37 + 77 = 139$$

Figure 2.18: Computation of the total walk count (twc) and of the walk complexity (wcx) for the methylcyclobutane graph using powers of the adjacency matrix.

(a) Graph G representing methylcyclobutane

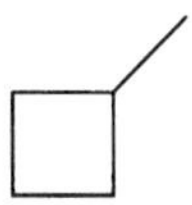

G

(b) Values of $awc_\ell(i)$ obtained by iterative summation over all neighbors, starting from vertex degrees, values of mwc_ℓ and awcs(i)

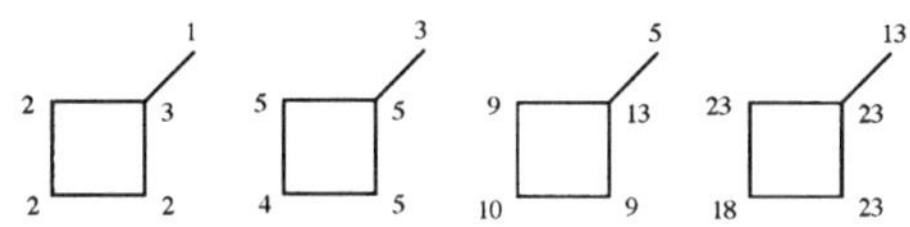

$mwc_1 = 10$ $mwc_2 = 22$ $mwc_3 = 46$ $mwc_4 = 100$

awcs(i):

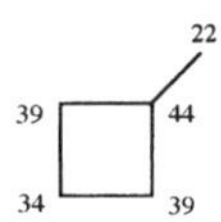

(c) Values of twc and wcx

twc = 22 + 44 + 39 + 34 + 39 = 178
wcx = 22 + 44 + 39 + 34 = 139

Figure 2.19: Computation of the total walk count (twc) and of the walk complexity (wcx) for the methylcyclobutane graph using the Morgan algorithm.

The molecular walk count of length ℓ (mwc_ℓ) then is:

$$mwc_\ell = 2^\ell \, V - \sum_{i=0}^{\ell-1} 2^{\ell-1-i} s_i \qquad (\ell \leq V) \tag{2.53}$$

and the total walk count (twc) is:

$$twc = mwc_V - 2\,V + \sum_{i=0}^{V-1} s_i \tag{2.54}$$

(ii) The unbranched cycloalkanes

$$mwc_\ell = 2^\ell \, V \tag{2.55}$$

$$\text{twc} = 2\text{V}\ (2^{\text{V}-1} - 1) \tag{2.56}$$

(iii) The regular graphs of degree d

$$\text{mwc}_\ell = \text{d}^\ell \text{V} \tag{2.57}$$

$$\text{twc} = \text{V d}\ (\text{d}^{\text{V}-1} - 1) / (\text{d} - 1) \tag{2.58}$$

Index twc does not take into account the graph's symmetry. The term labyrinthicity was conceived for what is measured by twc: finding yourself placed inside a labyrinth and trying to find a way out, you cannot profit from the symmetry which may be built into the labyrinth. The difficulty of the task increases as the size, branching, cyclicity, etc. of the labyrinth increase. To profit from symmetry you must be able to perceive it, and this requires a total view, a view from the outside. Thus, those who argue that a measure of complexity should take into account the symmetry, may be better served by a modification of twc (modified by symmetry) which is dubbed the walk complexity, denoted by wcx, and obtained simply from equation (2.49) [58]: in the summation of the awcs(i) values over the vertices all those vertices are left out that are equivalent by symmetry to any vertex already considered:

$$\text{wcx} = \sum_{\text{i}} \text{awcs (i)} \tag{2.59}$$

where i is the first integer from each set of integers denoting vertices in an equivalence class by symmetry. The calculation of wcx is likewise shown in Figures 2.18 and 2.19.

Closed formulas can be given for wcx of *n*-alkanes:

$$\text{wcx} = \left(\text{mwc}_\text{V} - 2\text{ V} + \sum_{\text{i}=0}^{\text{V}-1} \text{s}_\text{i} \right) / 2 \qquad \text{if V is even} \tag{2.60}$$

and for cycloalkanes:

$$\text{wcx} = 2\ (2^{\text{V}-1} - 1) \tag{2.61}$$

Index wcx being a variant of twc can be made sensitive to multiple bonds and heteroatoms as can the parent index [58]. The values of twc and wcx are obtained using the program MORGAN [89].

2.3.9 The Principal Eigenvalue

The principal eigenvalue, denoted λ_1, of a graph's adjacency matrix (the numerically largest component in the graph's spectrum) is essentially an averaged extended vertex degree, where the whole structure is taken into account (as opposed to local structural features). Thus, it is bounded by the lowest and highest vertex degree present in the graph. It is intimately related to the number of walks, as shown by Cvetković and Gutman [84] and ourselves [85]. Index λ_1 has several attractive properties as a complexity index: (i) it is very sensitive to details of branching, cyclicity, multiple edges, heteroatoms, all in the correct sense, i.e., it increases with an increase in the counts of these structural features, and (ii) values of λ_1 are easily obtained from any computer program performing matrix algebra. Nevertheless, index λ_1 has not been systematically investigated as a complexity index. The reasons for this may be found in two obvious inherent disadvantages immediately following from its nature: (i) it cannot discriminate regular graphs of degree d. In a regular graph of degree d all vertices have the same degree d, so its λ_1 value is d; (ii) it is almost independent of size, so that for most simple graphs of saturated hydrocarbons the λ_1 values are squeezed into the narrow range between about 1.8 and 3.0, which results in considerable degeneracy.

For the *n*-alkanes, λ_1 is given as [84,86]:

$$\lambda_1 = 2\cos[\pi/(V+1)] \tag{2.62}$$

which for increasing V converges toward 2, while for cycloalkanes λ_1 trivially equals 2.0.

2.3.10 Miscellaneous Approaches

A few other approaches to molecular complexity cannot be discussed here but should at least be mentioned.

Gamziani [49] proposed a molecular complexity index which is based on the number of reactions a compound can formally undergo. These reactions are constructed using graph-theoretical tools.

A quantum-chemical approach by Luzanov [50] does not explicitly refer to graph theory.

Whitlock [87] measured molecular size as the count of bonds, and molecular complexity in terms of the count of the structural features such as rings, multiple bonds, heteroatoms, and stereocenters. These numbers were weighted using intuitively found factors and then summed. Remarkably, this measure completely ignores adjacency information (the interaction of elements in a system, the focus of the present article), but nevertheless was found useful in visualizing the complexity changes occurring in the total synthesis of natural products. For this purpose otherwise Bertz's indices were used [1, 54, 88].

2.4 Comparison between the Reviewed Complexity Indices for Selected Molecular Graphs

We compared the reviewed complexity indices on five sets of selected molecular graphs. According to the most fundamental criterion for a complexity index, the homology principle, the index value should increase with increasing graph size. Therefore the first set of selected structures is the set of *n*-alkanes. The *n*-alkane trees up to V = 10 and their complexity indices are given in Table 2.1.

All of the considered indices increase with increasing *n*-alkane size. The Minoli and Minoli–Bonchev indices increase with size because all the structural parameters included in their formulas (V, E, $\Sigma P_\ell, \Lambda_p$) increase with increasing size of the *n*-alkanes. BT and BI can analytically be shown to increase for increasing V using equations (2.10) and (2.14).

The N_s and N_t numbers for *n*-alkanes steadily increase with the number of vertices, as described by equations (2.22) and (2.23). Each *n*-alkane in the series has an N_t number made up from the N_t number of the preceding *n*-alkane and its own size in terms of the number of vertices. For example, the N_t value of nonane (45) is equal to N_t(octane) + 9. Both Bonchev indices TC and TC1 increase with the size of *n*-alkanes since they are both sums of contributions of all connected subgraphs.

The Randić complexity index increases in groups of two (propane and butane; pentane and hexane; heptane and octane, etc.) reflecting the number of vertices in different environments. For example, in propane there are two end vertices (one type of vertices) and one central vertex (the second type of vertices). In butane there are again two end vertices and two central vertices. Thus, propane and butane possess Randić complexity indices which differ only slightly. The same is encountered for the duo pentane and hexane; their RCI values also differ slightly. However, there is a considerable jump between their RCI and that of the preceding duo propane and butane. This is because pentane and hexane possess three different types of vertices.

The M_1 number increases with the size of the *n*-alkane in steps of 4, M_1(V-alkane) = M_1(V–1-alkane) + 4. This is so because a newly added atom to the *n*-alkane gets the degree one but the previously terminal atom gets the degree two. The M_2 number also increases with size in steps of 4,

Table 2.1: Complexity indices of linear trees depicting carbon skeletons of *n*-alkanes with up to ten carbon atoms.

n-Alkane tree	Complexity index#																	
	MI	MB	BT	BI	N_s	N_t	TC	TC1	RCI	M_1	M_2	TM_1	TM_1^*	TM_2	TM_2^*	twc	wcx	λ_1
	0.7	0.7	0	–	2	3	4	2	1.5	2	1	4	2	1	1	2	1	1.00000
	3.6	4.8	2.8	0.0	3	6	14	8	5.3	6	4	22	10	8	6	10	7	1.41421
	10.3	17.1	8.8	2.0	4	10	32	20	6.4	10	8	56	28	28	19	32	16	1.61803
	22.2	44.4	18.5	7.5	5	15	60	40	11.4	14	12	110	60	64	44	88	56	1.73205
	40.9	95.5	32.2	12.0	6	21	100	70	12.1	18	16	188	110	120	85	222	111	1.80194
	67.8	180.9	50.4	19.2	7	28	154	112	17.7	22	20	294	182	200	146	536	320	1.84776
	104.5	313.6	73.1	25.0	8	36	224	168	18.0	26	24	432	280	308	231	1254	617	1.87939
	152.5	508.2	100.6	33.3	9	45	312	240	23.8	30	28	606	408	448	344	2878	1664	1.90211
	213.2	781.6	133.1	40.0	10	55	420	330	24.0	34	32	820	570	624	489	6500	3250	1.91899

MI = Minoli index; MB = Minoli–Bonchev index; BT = Bonchev–Trinajstić index; BI = Bertz index; N_s, N_t = number of kinds and total number of connected subgraphs; TC, TC1 = Bonchev indices; RCI = Randić complexity index; M_1, M_2, TM_1, TM_1^*, TM_2, TM_2^* = Zagreb indices; twc = total walk count; wcx = symmetry-modified walk count = walk complexity; λ_1 = principal eigenvalue.

that is, M_2(V-alkane) = M_2(V–1-alkane) + 4 (for V > 3). The above argument applies also here: A newly added atom to the *n*-alkane forms a bond with the previously terminal atom and this bond gets the weight 2, while the next, previously terminal bond, gets now the weight four. The Zagreb indices TM_1, TM_1^*, TM_2 and TM_2^* also increase with size since they are the sums of M_1 (M_2) for all connected subgraphs. Thus, complexity indices (N_t, TC, TC1, TM_1, TM_1^*, TM_2 and TM_2^*) that depend on the number of connected subgraphs behave similarly.

On going from an *n*-alkane to the next higher *n*-alkane, index twc increases for two reasons: (1) another higher matrix power is included; and (2) the sum of the entries in each matrix power is larger for the higher *n*-alkane, i.e., the adjacency matrix of the higher *n*-alkane contains two more entries "1." The same applies to the increase of the wcx values.

The principal eigenvalue λ_1 increases more and more slowly for increasing V to approach the limit 2.0 for an *n*-alkane of infinite length.

To probe the influence of branching (which unfortunately cannot be isolated from the influence of symmetry), we chose as a second set of compounds the five isomeric acyclic hexanes. Their complexity indices are given in Table 2.2. The Minoli index is not given for the isomeric hexanes since its value is constant throughout this series (40.9).

The ordering of hexane isomers by the considered complexity indices is given in Table 2.3.

Branching is reflected in vertex degrees higher than two; trees with vertex degrees one and two only are not branched. The symmetry may be measured in terms of the number of vertices not equivalent by symmetry. The balance between branching and symmetry determines the complexity order of the hexane isomers. Ten symmetry-independent indices (N_t, TC, TC1, M_2, TM_1, TM_1^*, TM_2, TM_2^*, twc, λ_1) support the following complexity order of the isomeric hexane trees: **I**<**II**<**III**<**IV**<**V**. The remaining two symmetry-independent indices behave as follows: the M_1 index differs from the above order by giving **II** = **III** instead of **II** < **III**, while MB gives the reverse ordering. So MB decreases with increasing branching; from its definition it follows that long paths, as present in unbranched alkanes, bring a large contribution to MB. The MB shares this undesirable feature with some other topological indices such as the Wiener index W, the Randić connectivity index χ, or the Hosoya index Z, all of which therefore cannot serve as complexity indices except (in the form of -MB etc.) among isomeric structures.

Among the symmetry-dependent indices, BT here behaves as does MB (leading to -BT being a possible complexity index for isomers rather than

Table 2.2: Isomeric hexane trees and their complexity indices

	Hexane tree				
Complexity index#					
	I	**II**	**III**	**IV**	**V**
MB	95.5	87.3	84.5	79.1	76.4
BT	32.2	28.6	27.4	23.5	22.6
BI	12.0	21.2	19.2	21.0	29.8
N_s	6	8	8	7	8
N_t	21	24	25	28	30
TC	100	127	136	164	181
TC1	70	88	94	112	122
RCI	12.1	21.9	17.8	9.8	19.5
M_1	18	20	20	22	24
M_2	16	18	19	21	22
TM_1	188	277	300	404	505
TM_1^*	110	146	158	196	222
TM_2	120	172	199	264	290
TM_2^*	85	114	125	156	173
twc	222	268	284	330	370
wcx	111	234	201	124	272
λ_1	1.80194	1.90211	1.93185	2.00000	2.07431

MB = Minoli–Bonchev index; BT = Bonchev–Trinajstić index; BI = Bertz index; N_s, N_t = number of kinds and total number of connected subgraphs; TC, TC1 = Bonchev indices; RCI = Randić complexity index; M_1, M_2, TM_1, TM_1^*, TM_2, TM_2^* = Zagreb indices; twc = total walk count; wcx = symmetry-modified walk count = walk complexity; λ_1 = principal eigenvalue.

BT), while BI, N_s and wcx at least agree in that for them **I** is the least and **V** the most complex structure among the five hexane isomers. A completely different ordering is obtained by the RCI index.

The next set of structures considered is the set of lower cycloalkanes. Their complexity indices are given in Table 2.4. All complexity indices given in Table 2.4 increase with the size of the cycloalkane. Index λ_1 is constant and equal to 2.0 throughout the series and therefore not given in Table 2.4.

Table 2.3: Ordering of isomeric hexane trees by the complexity indices reported in Table 2.2

Complexity index	Complexity ordering from the least to the most complex structure
-MB	I < II < III < IV < V
-BT	I < II < III < IV < V
BI	I < III < IV < II < V
N_S	I < IV < II = III = V
N_t	I < II < III < IV < V
TC	I < II < III < IV < V
TC1	I < II < III < IV < V
RCI	IV < I < III < V < II
M_1	I < II = III < IV < V
M_2	I < II < III < IV < V
TM_1	I < II < III < IV < V
TM_1^*	I < II < III < IV < V
TM_2	I < II < III < IV < V
TM_2^*	I < II < III < IV < V
twc	I < II < III < IV < V
wcx	I < IV < III < II < V
λ_1	I < II < III < IV < V

The Minoli index and Minoli–Bonchev index increase with the size of cycloalkanes because their structural parameters ($V, E, \Sigma P_\ell, \Lambda_p$) increase with increasing size of cycloalkanes. Both Minoli and Minoli–Bonchev indices are bigger for cycloalkanes than for *n*-alkanes, because the $E, \Sigma P_\ell, \Lambda_p$ parameters are bigger for a cyclic structure than for the corresponding chain. For example, in butane and cyclobutane the parameters are: V(butane) = V(cyclobutane) = 4, E(butane) = 3, E(cyclobutane) = 4, ΣP_ℓ (butane) = 6, ΣP_ℓ (cyclobutane) = 12, Λ_p (butane) = 10, Λ_p (cyclobutane) = 24.

Equation (2.11) for BT(cycloalkane) is easily recognized as monotonically increasing with V. The difference BT(*n*-alkane)–BT(cycloalkane) is positive for each V, as seen by comparison of Table 2.1 and Table 2.4. The same can be easily derived from equations (2.10) and (2.11). Thus, the value of BT for a given cycloalkane is smaller than that of the corresponding *n*-alkane; the BT does not obey postulate BeZ2. This may be qualitatively understood by observing that upon closure of an *n*-alkane to form the cycloalkane all distances larger than V/2 disappear and are replaced by distances between 1 and V/2. So, in the cycloalkane, the

Table 2.4: Complexity indices of cycles depicting carbon skeletons of cycloalkanes with up to ten carbon atoms

Cycle	Complexity index#																
	MI	MB	BT	BI	t(G)	N_s	N_t	TC	TC1	RCI	M_1=M_2	TM_1	TM_1^*	TM_2	TM_2^*	twc	wcx
	9	13.5	0	4.8	3	4	10	42	24	4.00	12	84	36	48	27	18	6
	24	48.0	5.5	8.0	4	5	17	88	56	4.50	16	176	88	112	68	56	14
	50	125.0	10.0	11.6	5	6	26	160	110	5.00	20	320	180	220	145	150	30
	90	270.0	22.8	15.5	6	7	37	264	192	5.25	24	528	324	384	270	372	62
	147	514.5	33.3	19.7	7	8	50	406	308	5.50	28	812	532	616	455	882	126
	224	896.0	54.6	24.0	8	9	65	592	464	5.63	32	1184	816	928	712	2032	254
	324	1458.0	72.0	28.5	9	10	82	828	666	5.75	36	1656	1188	1332	1053	4590	510
	450	2250.0	102.6	33.2	10	11	101	1120	920	5.81	40	2240	1660	1840	1490	10220	1022

MI = Minoli index; MB = Minoli–Bonchev index; BT = Bonchev–Trinajstić index; BI = Bertz index; t(G) = the number of spanning trees; N_s, N_t = number of kinds and total number of connected subgraphs; TC, TC1 = Bonchev indices; RCI = Randić complexity index; M_1, M_2, TM_1, TM_1^*, TM_2, TM_2^* = Zagreb indices; twc = total walk count; wcx = symmetry-modified walk count = walk complexity.

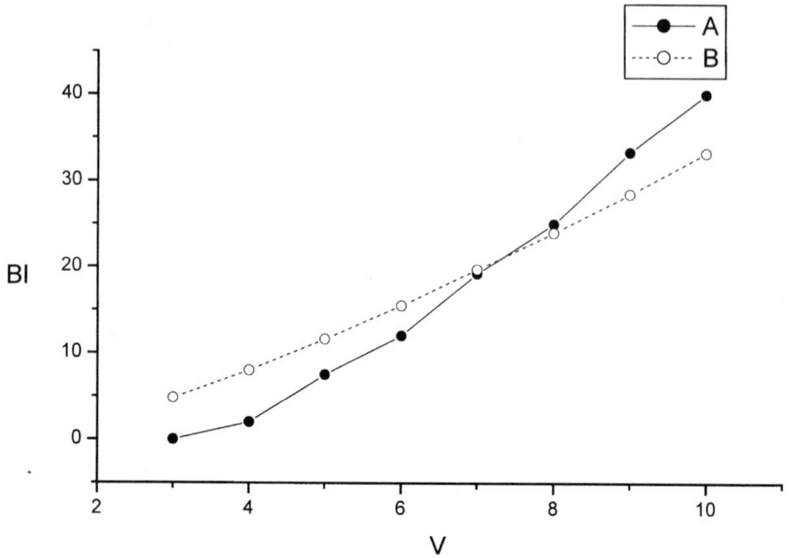

Figure 2.20: A plot of BI *versus* V for *n*-alkanes (A) and unbranched cycloalkanes (B).

distribution of distances is more uniform than in the *n*-alkane, the sum (to be subtracted) of log terms in equation (2.9) is larger in the cycloalkane.

The Bertz index BI for unbranched cycloalkanes increases with increasing size according to equation (2.15). On comparing the BI values from Tables 2.1 (*n*-alkanes) and 2.4 (cycloalkanes), it is seen that the two functions BI(V) cross at a V value between 7 and 8 (see Figure 2.20), i.e., there is no consistent picture whether the *n*-alkane or the cycloalkane of V vertices is more complex according to BI. Thus, BI does not satisfy criterion BeZ2 for an index of intrinsic complexity. The BI index heavily depends on symmetry, and therefore in Bertz's terminology rather is an index of extrinsic complexity, as are the other symmetry-dominated indices RCI and wcx, which deliberately violate criterion BeZ2 (see below).

The number of spanning trees t(G), and N_s and N_t increase with the size of cycloalkanes as described by equations (2.16), (2.24) and (2.25). Comparison of these formulas with those for the corresponding *n*-alkanes, given in equations (2.22) and (2.23), shows that the numbers for cycloalkanes are always larger.

The TC and TC1 indices increase with the size of cycloalkanes since they are made up of contributions from all connected subgraphs that belong to a given cycloalkane graph. The TC and TC1 indices are always bigger for cycloalkanes than for the corresponding *n*-alkanes because the number of subgraphs is always bigger for a given cycloalkane than for the corresponding *n*-alkane. For example, cycloheptane has in total 50 subgraphs, while *n*-heptane has only 28 subgraphs.

The Randić complexity index increases rather slowly for the cycloalkanes, as seen from equation (2.33), because all vertices in a cycle are equivalent by symmetry. For the same reason RCI for a given unbranched cycloalkane is always smaller than that for the corresponding *n*-alkane. This is easily seen if we compare the Randić complexity indices for *n*-pentane and cyclopentane. While *n*-pentane has three symmetry non-equivalent atoms:

$$\begin{aligned}\text{RCI } (n\text{-pentane}) = \; & 1 + 2/2 + 2/4 + 2/8 + 1/16 \\ & + 2 + 3/2 + 2/4 + 1/8 \\ & + 2 + 4/2 + 2/4 = 11.44\end{aligned}$$

cyclopentane possesses only one such atom:

$$\text{RCI (cyclopentane)} = 2 + 4/2 + 4/4 = 5.$$

The M_1 and M_2 indices are identical for a cycloalkane and increase in steps of four, $M_{1,2}$ (V-cycloalkane) = $M_{1,2}$ (V – 1-cycloalkane) + 4. This is so because in cycloalkanes the squares of the vertex degrees are equal to the edge weights, 4. The difference between the M indices for *n*-alkanes and cycloalkanes is given by M_1 (cycloalkane) – M_1 (*n*-alkane) = 6 for all V, and M_2 (cycloalkane) – M_2 (*n*-alkane) = 8 for $V \geq 3$.

The TM_1, TM_1^*, TM_2, TM_2^* indices increase with the size of cycloalkanes because their computation is based on the number of subgraphs. Since the number of subgraphs increases with increasing size, these indices must also increase with the size of cycloalkanes. The TM_1, TM_1^*, TM_2, TM_2^* indices are bigger for cycloalkanes than for *n*-alkanes because the total number of subgraphs for a given cycloalkane is bigger than for the corresponding *n*-alkane.

The reasons for the increase of twc and wcx with increasing size of cycloalkanes are the same two as in the case of *n*-alkanes (see above). Index twc is always bigger for a given cycloalkane than for the corresponding *n*-alkane since the cycloalkane has one additional bond, i.e., its adjacency matrix contains two additional entries "1." The wcx index is always smaller for a given cycloalkane than for the corresponding *n*-alkane since the number of symmetry non-equivalent vertices in unbranched cycles is one, while in the corresponding *n*-alkanes the number

Table 2.5: All connected molecular graphs with four vertices and their complexity indices

	Molecular graph with four vertices					

Complexity index#	**VI**	**VII**	**VIII**	**IX**	**X**	**XI**
MI	10.3	10.3	22.0	24.0	42.2	72.0
MB	17.1	15.4	40.0	48.0	86.7	158.4
BT	8.8	6.0	5.5	5.5	3.9	0.0
BI	2.0	4.8	19.2	8.0	36.0	43.0
t(G)	1	1	3	4	8	16
N_s	4	4	7	5	9	10
N_t	10	11	18	17	33	64
TC	32	39	98	88	254	648
TC1	20	24	60	56	152	372
RCI	6.4	7.5	13.8	4.5	12.0	7.5
M_1	10	12	18	16	26	36
M_2	8	9	19	16	33	54
TM_1	56	87	230	176	670	1944
TM_1^*	28	36	100	88	280	756
TM_2	28	36	145	112	504	1674
TM_2^*	19	24	80	68	252	762
twc	32	36	64	56	102	156
wcx	16	22	47	14	51	39
λ_1	1.61803	1.73205	2.17009	2.00000	2.56155	3.00000

MI = Minoli index; MB = Minoli–Bonchev index; BT = Bonchev–Trinajstić index; BI = Bertz index; t(G) = the number of spanning trees; N_s, N_t = number of kinds and total number of connected subgraphs; TC, TC1 = Bonchev indices; RCI = Randić complexity index; M_1, M_2, TM_1, TM_1^*, TM_2, TM_2^* = Zagreb indices; twc = total walk count; wcx = symmetry-modified walk count = walk complexity; λ_1 = principal eigenvalue.

of symmetry non-equivalent vertices increases with increasing size. Index λ_1 equals 2.0 for unbranched cycloalkanes and thus is larger than for *n*-alkanes.

In Table 2.5 we give all connected simple graphs with four vertices and their complexity indices, and their complexity orderings are given in Table 2.6. These graphs may be used to represent *n*-butane (**VI**), isobutane (**VII**), methylcyclopropane (**VIII**), cyclobutane (**IX**), bicyclo[1.1.0]butane (**X**) and tetrahedrane (**XI**).

Table 2.6: Ordering of all connected molecular graphs with four vertices by the complexity indices reported in Table 2.5

Complexity index	Complexity ordering from the least to the most complex structure
MI	**VI = VII < VIII < IX < X < XI**
MB	**VII < VI< VIII < IX < X < XI**
-BT	**VI < VII < VIII = IX < X < XI**
BI	**VI < VII < IX < VIII < X < XI**
t(G)	**VI = VII < VIII < IX < X < XI**
N_S	**VI = VII < IX < VIII < X < XI**
N_t	**VI < VII < IX < VIII < X < XI**
TC	**VI < VII < IX < VIII < X < XI**
TC1	**VI < VII < IX < VIII < X < XI**
RCI	**IX < VI < VII = XI < X < VIII**
M_1	**VI < VII < IX < VIII < X < XI**
M_2	**VI < VII < IX < VIII < X < XI**
TM_1	**VI < VII < IX < VIII < X < XI**
TM_1^*	**VI < VII < IX < VIII < X < XI**
TM_2	**VI < VII < IX < VIII < X < XI**
TM_2^*	**VI < VII < IX < VIII < X < XI**
twc	**VI < VII < IX < VIII < X < XI**
wcx	**IX < VI < VII < XI < VIII < X**
λ_1	**VI < VII < IX < VIII < X < XI**

In the case of all connected molecular graphs with four vertices, three structural characteristics are present: branching, cyclicity and symmetry. If we consider only branching, the following branching order is obtained: (i) acyclic structures **VI** < **VII**; and (ii) cyclic structures **IX** < **VIII** < **X** < **XI**. The first order was produced by the following set of complexity indices: BI, N_t, TC, TC1, RCI, M_1, M_2, TM_1, TM_1^*, TM_2, TM_2^*, twc, wcx, λ_1; and the second order was produced by the same set of indices except RCI and wcx, plus the N_s index. If we consider only cyclicity in terms of either closing a cycle or increasing the number of cycles, then the following cyclicity order is obtained: (i) closing a cycle **VI** < **IX**; and (ii) increasing the number of cycles **IX** (one cycle) < **X** (two cycles) < **XI** (three cycles). Both orders were followed by most of the considered complexity indices. Only those complexity indices that are strongly symmetry-dependent (BT, RCI, wcx) produced different orderings. If we consider only symmetry, then the order according to the number of symmetry non-equivalent atoms is **IX** = **XI** < **VI** = **VII** = **X** < **VIII**. Not a single index among the complexity indices gives this order, simply because complexity indices are *not* symmetry indices. The number of symmetry non-equivalent atoms is, of course, exactly reproduced by TOPSYM [74].

The balance between branching, cyclicity and symmetry should lead to the complexity order of considered graphs with four vertices. The following complexity order is supported by twelve indices (BI, N_t, TC, TC1, M_1, M_2, TM_1, TM_1^*, TM_2, TM_2^*, twc, λ_1): **VI** < **VII** < **IX** < **VIII** < **X** < **XI**. N_s differs only in not being able to differentiate **VI** and **VII**. BT gives numerically almost the reverse ordering, that is, –BT results in an ordering that differs from the above only in not differentiating **VIII** and **IX**. MI and t(G) cannot differentiate **VI** and **VII** and reverse **VIII** and **IX**. MB reverses **VIII/IX** and **VI/VII**.

The strongly symmetry-dependent indices RCI and wcx as expected result in orderings which are very different from the above and somewhat similar to one another.

In the last set of compounds we tried to isolate the influence of symmetry. A series of compounds having identical size, branching and cyclicity is provided by the saturated $(CH)_8$ isomers cubane, cuneane, octabisvalene (**XII–XIV**) and their hypothetical graph-theoretically non-planar isomers möbiuscubane and möbiuscuneane (**XV** and **XVI**). The corresponding complexity index values are given in Table 2.7 together with values of three measures of symmetry.

Apart from the number of topologically different vertices (vertex classes, first line), these are the numbers of topologically different edges (edge classes, second line) and of topologically different pairs of vertices (pair classes, third line). Gratifyingly, these measures agree as to the ordering of **XII** – **XVI** except for a few cases of non-discriminance. The orderings obtained by the symmetry indices and by the complexity indices appear in Table 2.8.

With other factors being constant throughout the series, the order of increasing complexity should be expected to be the order of decreasing symmetry. The four strongly symmetry-dependent indices BI, N_s, RCI and wcx in fact live up to this expectation in that they order these isomers exactly or nearly as do the symmetry indices.

Most other indices are able to discriminate between the isomers. Exceptions are M_1, M_2, twc and λ_1, that is, the definitely not symmetry-dependent indices.

A group of indices (MI, MB, t(G), N_t, TC, TC1) results in an identical order different from the above. These are the indices depending on the number of paths, spanning trees or connected subgraphs. They agree in

Table 2.7: Symmetry indices and complexity indices of graphs representing the saturated $(CH)_8$ isomers

Index[#]	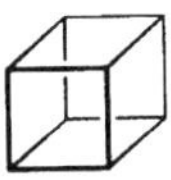 **XII**	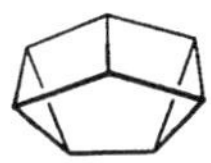 **XIII**	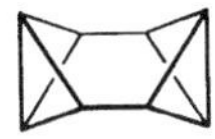 **XIV**	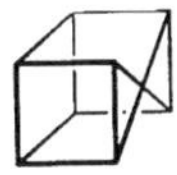 **XV**	 **XVI**
vertex cl	1	3	2	1	3
edge cl	1	5	3	2	3
pair cl	3	10	7	4	7
MI	2131.2	2006.4	1777.4	2169.6	2102.4
MB	10425.6	9820.8	8697.6	10656.0	10353.6
BT	40.566	36.283	42.857	27.586	27.586
BI	110.039	176.078	156.078	132.078	164.039
t(G)	384	336	256	392	363
N_S	64	195	117	102	148
N_t	2441	2245	1860	2472	2356
TC	51384	47262	39120	52008	49578
TC1	32712	30300	25376	33064	31650
RCI	10.125	31.125	19.875	10.500	31.500
M_1	72	72	72	72	72
M_2	108	108	108	108	108
twc	26232	26232	26232	26232	26232
wcx	3279	9837	6558	3279	9837

Symmetry indices: vertex cl = vertex classes; edge cl = edge classes; pair cl = pair classes. Complexity indices: MI = Minoli Index; MB = Minoli–Bonchev index; BT = Bonchev–Trinajstić index; BI = Bertz index; t(G) = number of spanning trees; N_S, N_t = number of kinds and total number of connected subgraphs; TC, TC1 = Bonchev indices; RCI = Randić complexity index; M_1, M_2, = Zagreb indices; twc = total walk count; wcx = symmetry-modified walk count = walk complexity.

attributing (by a considerable margin) the lowest complexity to octabisvalene (**XIV**). In fact, octabisvalene has far fewer paths, spanning trees and connected subgraphs than the other isomers. This is a consequence of its unique structure made of two halves connected by only two edges. Cutting these two edges would result in decomposition of octabisvalene into two pieces, whereas all other isomers require three cuts to be decomposed; **XIV** is said to be 2-connected, whereas its isomers are 3-connected. Thus, for octabisvalene there is a combination of edges (the two special ones) which when removed

Table 2.8: Ordering of saturated $(CH)_8$ isomers by the symmetry indices and by the complexity indices reported in Table 2.7

Symmetry indices	
vertex cl	**XII = XV < XIV < XVI = XIII**
edge cl	**XII < XV < XIV = XVI < XIII**
pair cl	**XII < XV < XIV = XVI < XIII**
Complexity indices	
MI	**XIV < XIII < XVI < XII < XV**
MB	**XIV < XIII < XVI < XII < XV**
-BT	**XIV < XII < XIII < XVI = XV**
BI	**XII < XV < XIV < XVI < XIII**
t(G)	**XIV < XIII < XVI < XII < XV**
N_S	**XII < XV < XIV < XVI < XIII**
N_t	**XIV < XIII < XVI < XII < XV**
TC	**XIV < XIII < XVI < XII < XV**
TC1	**XIV < XIII < XVI < XII < XV**
RCI	**XII < XV < XIV < XVI < XIII**
M_1	constant
M_2	constant
twc	constant
wcx	**XII = XV < XIV < XVI = XIII**
λ_1	constant

results in disconnected subgraphs, whereas for the other isomers such a dangerous edge combination does not exist.

The situation is similar for the paths. In **XIV** all of the longer paths have to pass through the two bottlenecks, whereas paths in the other isomers do not suffer such restrictions.

To obtain a spanning tree from one of **XII–XVI**, five edges have to be cut. In **XIV** each combination of edges which includes the two special ones does not result in a spanning tree, whereas no such edge combination exists for the other isomers.

Another consequence of octabisvalene's special structure is that its vertices are on average further apart than the vertices in the other isomers: **XIV** has 6 pairs of vertices at (the maximal) distance 3, whereas **XII** has 4 such pairs, **XIII** has 2, **XV** and **XVI** have none. Based on distances are indices such as BT or the Wiener index W. W simply measures the distances, while BT measures the variance in the distribution of distances. For both measures **XIV** assumes the maximum, whereas **XV** and **XVI** assume the minimum value among the isomers. In fact BT and W produce exactly the same order of the $(CH)_8$ isomers, which is similar to the path- or subgraph-induced order in that **XIV** is the least, whereas **XV** (together with **XVI**) is the most complex structure.

We have thus identified another factor contributing to complexity: Its 2-connectedness makes octabisvalene less complex than its 3-connected isomers, if the complexity is measured in terms of paths, connected subgraphs, spanning trees or distances.

2.4.1 Practical Considerations

We have compared the numbers and orderings obtained from several complexity indices. However, there are some practical aspects associated with complexity indices that will be briefly discussed here. It was already mentioned (section 2.3.3.4) that t(G) tends to give large numbers even for not-so-large graphs. In fact, this is a problem with most complexity indices. Tables 2.1, 2.2, 2.4, 2.5 and 2.7 demonstrate that also MI, MB, N_t (and its derivatives TC, TC1, TM_1, TM_1^*, TM_2, TM_2^*), twc and wcx grow very fast for (poly)cyclic graphs.

Thus, the number of paths increases quickly with increasing size and cyclicity, subgraphs generally are more numerous than spanning trees, and the number of walks is largest of all those considered here, as described by the formulas in section 2.3. The essence of these formulas, the order of the increase of index values as a function of V, is compiled in Table 2.9.

The sheer size of the integer numbers, on the other hand, is more or less a computer technical problem. There are the problems of data overflow and of precision. As to numerical precision, the situation is the same as for real numbers: Any digits given beyond a certain precision should not be trusted.

Of more concern is the computational effort required to obtain the numerical values. This effort is not primarily dependent on the magnitude of the resulting number, but on the difficulty of the procedure. In Table 2.10 numerical values of selected graph invariants for the dodecahedrane graph are given together with the CPU times used to obtain these numbers using our programs on our computer, an SG Indigo workstation, 150 MHz, R5000 coprocessor. The short CPU times for t(G) and twc result from the fact that a mathematical formula is known for the former and a simple procedure (nothing but addition steps) is valid for the latter. In both cases the count only is required, not a knowledge of the individual spanning trees or walks. In contrast, for paths and connected subgraphs no general mathematical formulas are known, so that paths and subgraphs have to be constructed individually by time-consuming path algorithms.

Table 2.9: Order of the increase of complexity indices for increasing V, in the series of *n*-alkanes and unbranched cycloalkanes

	***n*-alkanes**	**Unbranched cycloalkanes**
MI	V^3	V^3
MB	V^4	V^4
BT	$V^2 \log_2 V$	$V^2 \log_2 V$
BI	$V \log_2 V$	$V \log_2 V$
t(G)	constant=1	V
N_s	V	V
N_t	V^2	V^2
TC	V^3	V^3
TC1	V^3	V^3
RCI	V	$\to 6$
M_1	V	V
M_2	V	V
twc	$V \cdot 2^{V-1}$	$V \cdot 2^{V-1}$
wcx	$V \cdot 2^{V-2}$	2^{V-1}
λ_1	$\to 2$	constant=2

2.5 Concluding Remarks

The notion of molecular complexity as conceived here is made up of several sub-notions such as size, branching, cyclicity, presence of multiple bonds or heteroatoms, symmetry, etc. In the absence of an experimental criterion we are free to weigh these contributions, and consequently various preferences result among individual researchers.

In this situation it is certainly impossible to identify the best complexity index. However, what we can do is at least to identify the weak points for each index. Thus,

- MI is insensitive to branching;
- MB and BT decrease with increasing branching, while increasing with increasing size, thus causing logical inconsistencies;
- t(G) is constant and equal to one for all acyclic graphs;
- N_s is rather degenerate, a computer program giving N_s values became available only recently [90];

Table 2.10: Numerical values for selected graph invariants for the dodecahedrane graph, and the CPU times required to obtain these numbers

Graph invariant	**Count**	**CPU time**	**Program**
Paths	125370	0.07 sec	PATHS
Spanning trees, t(G)	5184000	< 0.01 sec	(NAG F02ABF)
Connected subgraphs, N_t	145168248	ca. 73 min	SUBGRAPH
Walks, twc	34867843980[a]	< 0.01 sec	MORGAN

[a]This number is arrived at independently by use of equation (2.58).

- a computer program giving N_t values now exists, but the procedure is exponential in problem size [90, 91];
- N_s, TC, TC1, TM_1, TM_1^*, TM_2, TM_2^* all depend on knowledge of all connected subgraphs, so they suffer from the same problem as does N_t;
- M_1 and M_2 are highly degenerate indices, that is, they cannot discriminate between many isomers;
- twc and λ_1 cannot discriminate between isomeric regular graphs;
- RCI and wcx emphasize (and may be said to overemphasize) the role of symmetry. BI has a strongly symmetry-dependent component and therefore does not result in a consistent ordering between cycloalkanes and the corresponding *n*-alkanes. To use these three indices requires explicit input of symmetry information, which in turn requires a reliable computer program for symmetry perception [59,74].

Acknowledgments

This work was supported by the Ministry of Science and Technology of the Republic of Croatia through Grant No. 00980606. We thank Dr Steven H. Bertz for useful discussions about his index by e-mail letters, Professors Alexandru T. Balaban, Subhash C. Basak, Danail Bonchev and Milan Randić for discussions about molecular complexity and approaches to quantify it, and Professor John N. Murrell FRS for discussions about possible experimental (spectroscopic) measures of complexity.

2.6 References

1. S.H. Bertz and W.F. Wright, The Graph Theory Approach to Synthetic Analysis: Definition and Application of Molecular Complexity and Synthetic Complexity, *Graph Theory Notes of New York*, **35**, 32–48 (1998).
2. M. Mitchell Waldrop, *Complexity*, Touchstone/Simon&Schuster, New York, 1992.
3. S.H. Strogatz, The Latest C-Theory, *Nature*, **387**, 771–772 (1997); C-theories are cybernetics (appearing in the 1960s), catastrophe theory (appearing in the 1970s), the chaos theory (appearing in the 1980s) and the complexity theory (appearing in the 1990s).
4. J.T. Bonner, *The Evolution of Complexity by Means of Natural Selection*, University Press, Princeton, 1988.
5. B. Kaye, *Chaos & Complexity*, VCH, Weinheim, 1993.
6. G.A. Cowan, D. Pines and D. Meltzer (Eds), *Complexity: Metaphors, Models and Reality*, Addison-Wesley Publ. Co., Reading, MA, 1994.
7. M. Gell-Mann, *The Quark and the Jaguar—Adventures in the Simple and the Complex*, Abacus, London, 1995.
8. A. Mowshowitz, Entropy and the Complexity of Graphs. I. An Index of the Relative Complexity of a Graph, *Bull. Math. Biophys.* **30**, 175–204 (1968).
9. R.A. Kerr, The Most Complex Magnetic Field, *Science*, **232**, 1603 (1986).
10. F. Papentin, On Order and Complexity. II. Application to Chemical and Biochemical Structures, *J. Theor. Biol.* **95**, 225–245 (1982).
11. R. Hinegardner and J. Engelberg, Biological Complexity, *J. Theor. Biol.* **104**, 7–20 (1983).
12. D.P. Bovet and P. Crescenzi, *Introduction to the Theory of Complexity*, Prentice Hall, New York, 1994.
13. P. Bürgisser, M. Clausen and M.A. Shokrollahi, *Algebraic Complexity Theory*, Springer, Berlin, 1997.
14. M. Crichton, *The Lost World* (inspired by Sir Arthur Conan Doyle's novel of the same title) Alfred A. Knopf, Inc., New York, 1995, p. 3. This is a sequel to the bestselling novel *Jurassic Park* by the same author (Alfred A. Knopf, Inc., New York, 1990).
15. G. Birkhoff, *Aesthetic Measure*, Harvard University Press, Cambridge, MA, 1933.
16. S.A. Kauffman, *The Origins of Order: Self-Organization and Selection in Evolution*, University Press, Oxford, 1993.

17. H. Moritz, *Science, Mind and the Universe*, Wichmann, Heidelberg, 1995.
18. (a) R. Badii and A. Politi, *Complexity: Hierarchical Structures and Scaling in Physics*, University Press, Cambridge, 1997; (b) P. Wagner, On Complexity, Grammars and the Inferring of Models, *Z. Naturforsch.* **48a**, 651–654 (1993).
19. *Webster's Ninth New Collegiate Dictionary*, Merriam-Webster Inc., Springfield, MA, 1985, p. 269.
20. E.g., *The American Heritage Dictionary*, Second College Edition, Houghton Mifflin Co., Boston, 1985, p. 302.
21. F. Harary, *Graph Theory*, second printing, Addison-Wesley, Reading, MA, 1972.
22. R.J. Wilson, *Introduction to Graph Theory*, Oliver and Boyd, Edinburgh, 1972.
23. D. Bonchev and D.H. Rouvray (Eds), *Chemical Graph Theory—Introduction and Fundamentals*, Gordon & Breach, New York, 1991.
24. N. Trinajstić, *Chemical Graph Theory*, 2nd revised edn, CRC Press, Boca Raton, FL, 1992.
25. D.H. Rouvray, Are the Concepts of Chemistry all Fuzzy?, in: *Concepts in Chemistry—A Contemporary Challenge*, D.H. Rouvray (Ed.), Wiley, New York, 1997, pp. 1–15.
26. V.I. Minkin, M.N. Glukhovtsev and B.Y. Simkin, *Aromaticity and Antiaromaticity*, Wiley, New York, 1994.
27. R. Dagani, Linked-Ring Molecules Wow Chemists with Their Elegance and Complexity, *Chem. Eng. News*, **72** (August 29), 28–34 (1994).
28. D. Bonchev, O. Mekenyan and N. Trinajstić, Topological Characterization of Cyclic Structures, *Int. J. Quantum Chem.* **17**, 845–893 (1980).
29. D. Bonchev, O.N. Temkin, and D. Kamenski, On the Complexity of Linear Reaction Mechanisms, *React. Kinet. Catal. Lett.* **15**, 119–124 (1980).
30. S.H. Bertz, The First General Index of Molecular Complexity, *J. Am. Chem. Soc.* **103**, 3599–3601 (1981).
31. S.H. Bertz, The Bond Graph, *J. Chem. Soc. Chem. Commun.* **1981**, 818–820.
32. S.H. Bertz, Convergence, Molecular Complexity and Synthetic Analysis, *J. Am. Chem. Soc.* **104**, 5801–5803 (1982).
33. S.H. Bertz, A Mathematical Model of Molecular Complexity, in: *Chemical Applications of Topology and Graph Theory*, R.B. King (Ed.), Elsevier, Amsterdam, 1983, pp. 206–221.

34. S.H. Bertz, On the Complexity of Graphs and Molecules, *Bull. Math. Biol.* **45**, 849–855 (1983).
35. O. Mekenyan, D. Bonchev and N. Trinajstić, Structural Complexity and Molecular Properties of Cyclic Systems with Acyclic Branches, *Croat. Chem. Acta*, **56**, 237–261 (1983).
36. S.H. Bertz and W.C. Herndon, The Similarity of Graphs and Molecules, in: *Artificial Intelligence Applications in Chemistry*, T.H. Pierce and B.A. Hohne (Eds), ACS, Washington, DC, 1986, pp. 169–175.
37. D. Bonchev and O.E. Polansky, On the Topological Complexity of Chemical Systems, in: *Graph Theory and Topology in Chemistry*, R.B. King and D.H. Rouvray (Eds), Elsevier, Amsterdam, 1987, pp. 126–158.
38. D. Bonchev, D. Kamenski and O.N. Temkin, Complexity Index for the Linear Mechanisms of Chemical Reactions, *J. Math. Chem.* **1**, 345–388 (1987).
39. S.C. Basak, Use of Molecular Complexity Indices in Predictive Pharmacology and Toxicology: A QSAR Approach, *Med. Sci. Res.* **15**, 605–609 (1987).
40. J.B. Hendrickson, P. Huang and G. Toczko, Molecular Complexity: A Simplified Formula Adapted to Individual Atoms, *J. Chem. Inf. Comput. Sci.* **27**, 63–67 (1987).
41. B.L. Mil'man, A Complexity Measure for Chemical Compounds, *J. Struct. Chem.* (English Translation), **29**, 957–960 (1988).
42. D. Bonchev, The Problems of Computing Molecular Complexity, in: *Computational Chemical Graph Theory*, D.H. Rouvray (Ed.), Nova Sci. Publ., Commack, NY, 1990, pp. 33–63.
43. S.H. Bertz and T.J. Sommer, Applications of Graph Theory to Synthesis Planning: Complexity, Reflexivity and Vulnerability, in: *Organic Synthesis: Theory and Applications*, Vol. II, JAI Press, Greenwich, CT, 1993, pp. 67–92.
44. G. Rücker and C. Rücker, Counts of All Walks as Atomic and Molecular Descriptors, *J. Chem. Inf. Comput. Sci.* **33**, 683–695 (1993).
45. E. Gordeeva, D. Bonchev, D. Kamenski and O.N. Temkin, Enumeration, Coding and Complexity of Linear Reaction Mechanisms, *J. Chem. Inf. Comput. Sci.* **34**, 436–445 (1994).
46. N. Trinajstić, S. Nikolić and Z. Mihalić, On the Complexity of Platonic Solids, *Bull. Chem. Technol. Macedonia*, **13**, 61–68 (1994).
47. L.B. Kier and B. Testa, Complexity and Emergence in Drug Research, *Adv. Drug Res.* **26**, 1–43 (1995).
48. D. Bonchev, Topological Order in Molecules. I. Molecular Branching Revisited. *J. Mol. Struct. (Theochem)*, **336**, 137–156 (1995).

49. (a) G.A. Gamziani, Methods for Evaluating Complexity of Organic Molecules and Reactions. I. Local Indices of Reaction Centers and Characteristics of Complexity of Organic Reactions, *Russ. J. Org. Chem.* (English Translation), **31**, 27–30 (1995); (b) G.A. Gamziani, Methods for Evaluating Complexity of Organic Molecules and Reactions. II. Formal Model of Structural Complexity of Organic Molecules, *Russ. J. Org. Chem.* (English Translation), **31**, 31–38 (1995).
50. A.V. Luzanov and E.N. Babich, Quantum-Chemical Quantification of Molecular Complexity and Chirality, *J. Mol. Struct. (Theochem)*, **333**, 279–290 (1995).
51. O.N. Temkin, A.V. Zeigarnik and D. Bonchev, *Chemical Reaction Networks: A Graph-Theoretical Approach*, CRC Press, Boca Raton, FL, 1996.
52. S. Nikolić, N. Trinajstić, A. Jurić, Z. Mihalić and G. Krilov, Complexity of Some Interesting (Chemical) Graphs, *Croat. Chem. Acta*, **69**, 883–897 (1996).
53. D. Bonchev and W.A. Seitz, The Concept of Complexity in Chemistry, in: *Concepts in Chemistry—A Contemporary Challenge*, D.H. Rouvray (Ed.), Wiley, New York, 1997, pp. 353–381.
54. S.H. Bertz and T.J. Sommer, Rigorous Mathematical Approaches to Strategic Bonds and Synthetic Analysis Based on Conceptually Simple New Complexity Indices, *Chem. Commun.* **1997**, 2409–2410; see also S.H. Bertz, Complexity of Molecules and Their Synthesis, chapter 3 in the present book.
55. D.G. Bonchev, Kolmogorov's Information, Shannon's Entropy, and Topological Complexity of Molecules, *Bulg. Chem. Commun.* **28**, 567–582 (1995).
56. D. Bonchev, Novel Indices for the Topological Complexity of Molecules, *SAR QSAR Environ. Res.* **7**, 23–43 (1997).
57. S. Nikolić, I.M. Tolić and N. Trinajstić, On the Complexity of Molecular Graphs, *MATCH-Commun. Math. Comput. Chem.* **40**, 187–201 (1999).
58. G. Rücker and C. Rücker, Walk Counts, Labyrinthicity and Complexity of Acyclic and Cyclic Graphs and Molecules, *J. Chem. Inf. Comput. Sci.* **40**, 99–106 (2000); see also I. Gutman, C. Rücker and G. Rücker, On Walks in Molecular Graphs, *J. Chem. Inf. Comput. Sci.* **41**, 739–745 (2001). In the former paper the index twc was defined as the half-sum of the respective mwc's. This definition reflected the fact that most walks in a graph occur in pairs. However, the palindromic self-

returning walks are self-reverse and consequently do not occur in pairs. Therefore, the earlier definition was modified in the latter paper by leaving out the multiplier 1/2. Thus, in the present article we used twc without the factor 1/2. Similarly, the factor 1/2 was removed from the definition of wcx.

59. M. Randić and D. Plavšić, On the Concept of Molecular Complexity, *Croat. Chem. Acta*, **75**, 107–116 (2002); see also M. Randić, On Complexity of Transitive Graphs Representing Degenerate Rearrangements, *Croat. Chem. Acta*, **74**, 683–705 (2001).
60. M. Randić and D. Plavšić, On Characterization of Molecular Complexity, *Int. J. Quantum Chem.* in press.
61. S.H. Bertz and C.M. Zamfirescu, New Complexity Indices Based on Edge Covers, *MATCH-Commun. Math. Comput. Chem.* **42**, 39–70 (2000).
62. S. Nikolić, I.M. Tolić, N. Trinajstić and I. Baučić, On the Zagreb Indices as Complexity Indices, *Croat. Chem. Acta*, **73**, 909–921 (2000).
63. J. Devillers and A.T. Balaban (Eds), *Topological Indices and Related Descriptors in QSAR and QSPR*, Gordon & Breach, Reading, UK, 1999. See the following chapters therein: (a) S.C. Basak, Information Theoretic Indices of Neighborhood Complexity and Their Applications, pp. 563–593; (b) D. Bonchev, Overall Connectivity and Topological Complexity. A New Tool for QSPR/QSAR, pp. 361–401. See also D. Bonchev and E. Gordeeva, Hierarchical Partially Ordered Sets Based on Topological Complexity, *MATCH-Commun. Math. Comput. Chem.* **42**, 85–117 (2000).
64. R. Merényi, J.F.M. Oth and G. Schröder, Die temperaturabhängigen NMR-Spektren von Tricyclo[3.3.2.0$^{4.6}$]decadien-(2.7), einigen seiner Derivate und von Tricyclo[3.3.2.0$^{4.6}$]decatrien-(2.7.9) (Bullvalen), *Chem. Ber.* **97**, 3150–3161 (1964); T.P. Živković, Bullvalene Reaction Graph, *Croat. Chem. Acta*, **69**, 215–222 (1996).
65. D. Minoli, Combinatorial Graph Complexity, *Atti Acad. Naz. Lincei Rend. Cl. Sci. Fis. Mat. Natur. (Ser. 8)*, **59**, 651–661 (1975).
66. C.E. Shannon and W. Weaver, *The Mathematical Theory of Communication*, University of Illinois Press, Urbana, IL, 1949.
67. D. Bonchev, *Information Theoretic Indices for Characterization of Chemical Structures*, Research Studies Press, Chichester, 1983.
68. (a) N. Rashevsky, Life, Information Theory, and Topology, *Bull. Math. Biophys.* **17**, 229–235 (1955); (b) N. Rashevsky, Contributions to Topological Biology: Some Considerations on the Primordial Graph

and on Some Possible Transformations, *Bull. Math. Biophys*. **18**, 113–128 (1956).

69. E. Trucco, A Note on the Information Content of Graphs, *Bull. Math. Biophys*. **18**, 129–135 (1956); E. Trucco, On the Information Content of Graphs: Compound Symbols; Different States for each Point, *Bull. Math. Biophys*. **18**, 237–253 (1956).
70. S.C. Basak, B.D. Gute and G.D. Grunwald, Estimation of the Normal Boiling Points of Haloalkanes Using Molecular Similarity, *Croat. Chem. Acta*, **69**, 1159–1173 (1996); S.C. Basak, G.D. Grunwald and G.J. Niemi, Use of Graph-Theoretic and Geometrical Molecular Descriptors in Structure-Activity Relationships, in: *From Chemical Topology to Three-Dimensional Geometry*, A.T. Balaban (Ed.), Plenum Press, New York, 1997, pp. 73–116; S.C. Basak, B.D. Gute, B. Lučić, S. Nikolić and N. Trinajstić, A Comparative QSAR Study of Benzamidines Complement-Inhibitory Activity and Benzene Derivatives Acute Toxicity, *Comput. Chem*. **24**, 181–191 (2000).
71. D. Bonchev and N. Trinajstić, Information Theory, Distance Matrix and Molecular Branching, *J. Chem. Phys*. **67**, 4517–4533 (1977).
72. L. Brillouin, *Science and Information Theory*, 2nd edn, Academic Press, New York, 1962.
73. (a) M. Randić, G.M. Brissey, R.B. Spencer and C.L. Wilkins, Search for All Self-Avoiding Paths for Molecular Graphs, *Comput. Chem*. **3**, 5–13 (1979); (b) J.V. Knop and N. Trinajstić, unpublished; (c) G. Rücker and C. Rücker, unpublished; see [83].
74. G. Rücker and C. Rücker, Computer Perception of Constitutional (Topological) Symmetry: TOPSYM, a Fast Algorithm for Partitioning Atoms and Pairwise Relations among Atoms into Equivalence Classes, *J. Chem. Inf. Comput. Sci.* **30**, 187–191 (1990); G. Rücker and C. Rücker, On Using the Adjacency Matrix Power Method for Perception of Symmetry and for Isomorphism Testing of Highly Intricate Graphs, *J. Chem. Inf. Comput. Sci.* **31**, 123–126 (1991); G. Rücker and C. Rücker, Isocodal and Isospectral Points, Edges, and Pairs in Graphs and How to Cope with Them in Computerized Symmetry Recognition, *J. Chem. Inf. Comput. Sci.* **31**, 422–427 (1991).
75. I. Gutman, R.B. Mallion and J.W. Essam, Counting the Spanning Trees of a Labelled Molecular-Graph, *Mol. Phys*. **50**, 859–877 (1983).
76. P.E. John, R.B. Mallion and I. Gutman, An Algorithm for Counting Spanning Trees in Labeled Molecular Graphs Homeomorphic to Cata-Condensed Systems, *J. Chem. Inf. Comput. Sci.* **38**, 108–112 (1998).

77. N. Trinajstić, D. Babić, S. Nikolić, D. Plavšić, D. Amić and Z. Mihalić, The Laplacian Matrix in Chemistry, *J. Chem. Inf. Comput. Sci.* **34**, 368–376 (1994).
78. D.M. Cvetković, M. Doob and H. Sachs, *Spectra of Graphs*, 3rd revised and enlarged edn, Johann Ambrosius Barth Verlag, Heidelberg, 1995, p. 27.
79. K. Balasubramanian, Laplacians of Fullerenes (C_{42}–C_{90}), *J. Phys. Chem.* **99**, 6509–6518 (1995).
80. M. Gordon and J.W Kennedy, The Graph-Like State of Matter. Part 2. LCGI Schemes for the Thermodynamics of Alkanes and the Theory of Inductive Inference, *J. Chem. Soc. Faraday Trans. II*, **69**, 484–504 (1973).
81. R.G.A. Bone and H.O. Villar, Exhaustive Enumeration of Molecular Substructures, *J. Comput. Chem.* **18**, 86–107 (1997).
82. G. Rücker and C. Rücker, Automatic Enumeration of All Connected Subgraphs, *MATCH-Commun. Math. Comput. Chem.* **41**, 145–149 (2000).
83. G. Rücker and C. Rücker, On Topological Indices, Boiling Points and Cycloalkanes, *J. Chem. Inf. Comput. Sci.* **39**, 788–802 (1999).
84. (a) L. Lovasz and J. Pelikan, On the Eigenvalues of Trees, *Period. Math. Hung.* **3**, 175–182 (1973). (b) D.M. Cvetković and I. Gutman, Note on Branching, *Croat. Chem. Acta*, **49**, 115–121 (1977).
85. C. Rücker and G. Rücker, Understanding the Properties of Isospectral Points and Pairs in Graphs. The Concept of Orthogonal Relation, *J. Math. Chem.* **9**, 207–238 (1992); C. Rücker and G. Rücker, Mathematical Relation between Extended Connectivity and Eigenvector Coefficients, *J. Chem. Inf. Comput. Sci.* **34**, 534–538 (1994).
86. D. Cvetković, P. Rowlinson and S. Simić, *Eigenspaces of Graphs*, University Press, Cambridge, 1997, p. 26.
87. H.W. Whitlock, On the Structure of Total Synthesis of Complex Natural Products, *J. Org. Chem.* **63**, 7982–7989 (1998).
88. (a) M. Chanon, R. Barone, C. Baralotto, M. Juillard and J.B. Hendrickson, Information Theory Description of Synthetic Strategies in the Polyquinane Series. The Holosynthon Concept, *Synthesis*, 1559–1583 (1998). (b) R. Barone and M. Chanon, A New and Simple Approach to Chemical Complexity. Application to the Synthesis of Natural Products, *J. Chem. Inf. Comput. Sci.* **41**, 269–272 (2001).
89. G. Rücker and C. Rücker, unpublished program MORGAN, see [44].
90. G. Rücker and C. Rücker, On Finding Non-isomorphic Connected Subgraphs and Distinct Molecular Substructures, *J. Chem. Inf. Comput. Sci.* **41**, 314–320, 865 (2001).

91. G. Rücker and C. Rücker, Substructure, Subgraph, and Walk Counts as Measures of the Complexity of Graphs and Molecules, *J. Chem. Inf. Comput. Sci.* **41**, 1457–1462 (2001).
92 C. Rücker, G. Rücker and M. Meringer, Exploring the Limits of Graph Invariant- and Spectrum-Based Discrimination of (Sub)structures, *J. Chem. Inf. Comput. Sci.* **42**, 640–650 (2002).

Chapter 3

COMPLEXITY OF MOLECULES AND THEIR SYNTHESIS

S.H. Bertz

Complexity Study Center, Mendham, NJ 07945, USA

3.1 What is complexity?

The subject of complexity is, well—complex. Since it is considered bad form to define a word in terms of words from which it is derived, we can consult the dictionary to see how the language experts define complexity. I have always disliked it when authors begin an article by saying "Webster's defines [subject] as ... ;" nevertheless, I could not resist my curiosity to see what my own timeworn Webster's [**1**] had to say: "complexity ... 1: the quality or state of being complex 2: something complex," which was my first thought, as expressed in the opening sentence. Next, I went to the root: "complex ... 1: a whole made up of complicated or *interrelated* parts," and finally "complicated ... 1: consisting of parts *intricately* combined 2: *difficult* to analyze, understand, or explain *syn* see COMPLEX." (The emphasis is mine throughout.) I had come full circle, back to the root. I thought the adverb 'intricately' might be the key, until I looked up "intricate ... 1: having many complexly *interrelated* parts or elements: COMPLICATED 2: *difficult* to resolve or analyze *syn* see COMPLEX." Thus, I found myself back at the root again. What I gleaned from this exercise was the importance of being *interrelated*—and that complexity is *difficult*! "Difficult" was not defined in terms of complexity, and it was with trepidation that I looked up "interrelated;" it did not say anything about being complex or intricate, so I was free of the tautology trap. One of the main themes of this chapter is the development of a definition of complexity in terms of the interactions among the parts of a system that will reflect how they are interrelated in a meaningful and useful way.

Before proceeding to quantification, we list the related entries in the *Dictionary of Scientific and Technical Terms* [2]: complex ("An assemblage of rocks that has been folded together, *intricately* mixed, involved, or otherwise complicated. A space which is represented as a union of simplices which intersect only on their faces. Composed of many ingredients. A group of associated ideas with strong emotional tones, which have been transferred from the conscious mind into the unconscious and which influence the personality."), complexation, complex chemical reaction, complex climatology, complex compound, complex conjugate, complex data type, complex declaration statement, complex degree of coherence, complex dune, complex fold, complex fraction, complex frequency, complex impedence, complexing, complexing agent, complex ion, complex low, complex notation, complex number, complexometric titration, complex permeability, complex permittivity, complex potential, complex reflector, complex relative attenuation, complex salt, complex sphere, complex target, complex tombolo, complex tone, complex unit, complex variable, complex velocity, complex wave.

It is interesting to see "intricately" again. The reason for presenting this information is to illustrate how the concept of complexity cuts across disciplines from chemistry and geology to psychology, physics and mathematics. More specific definitions of complexity were recently given by Bonchev [3]:

Definition 1: The higher the information content of a system, the more complex the system.

Definition 2: The larger the size of the system, and the higher the degree of its connectedness, the more structurally complex the system.

Definition 3: Molecular complexity increases with the number of rings, multiple bonds, and branches, as well as with molecular size, and with the variety in elemental composition and functional groups, and it decreases with symmetry.

Definition 4A: The general complexity of any system increases with the increase in the complexity of its size, topology, specific nature, and metric, and with the decrease in its symmetry.

Definition 4B: The topological complexity of any system increases with the increase in its connectedness and adjacency, with the presence of more complex connectivity patterns (cycles and branches), with the larger number of distances, paths, and walks, with the larger number of subgraphs of increasing size, and with the lower topological symmetry.

Definition 5: The higher the connectivity of the molecular graph and its connected subgraphs, the more topologically complex the molecule.

Information, connectedness and symmetry are recurrent themes, which will be discussed in more detail below. It was pointed out in the first publications on molecular complexity [4–8] that an index of it should increase with the number of each factor that contributes to the complexity of a molecule, e.g., atoms, bonds, multiple bonds, branches, rings, lone pairs of electrons and heteroatoms. Minoli's seminal paper [9] on combinatorial graph complexity was the basis of Definition 2 above, and we précis it in his own words. "A complexity function should enjoy the following properties:

(1) It should be monotonically increasing on the number of vertices of the graph;
(2) It should be monotonically increasing on the number of edges of the graph;
(3) It should reflect the degree of connectedness of the graph;
(4) It should satisfy our intuitive 'feel' of complexity, assigning a high number to a graph which 'looks' complicated, and vice versa."

Minoli's graph complexity function $\chi(G)$ is given in equation 3.1, where n is the number of vertices, e is the number of edges, and σ_{ij} is the number of proper paths between vertices i and j. (Formal definitions are given in the next section.) Selected values of $\chi(G)$ are summarized in Table 3.1. It is easy to verify that equation 1 satisfies the first two criteria, the third depends on the definition of connectedness, and the last one is a matter of opinion. In some cases it is easy to decide which of two graphs is more complex based on intuition, e.g., K_4 (tetrahedrane) > P_4 (butane), but other cases are not so clear-cut. What about C_4 (cyclobutane) and $K_{1,3}$

Table 3.1: Selected topological indices for some homologous series[a]

G	η	$C(\eta)$	$\chi(G)$	T_{span}	$T_S(G)$	$T_T(G)$
P_1	0	—[b]	0	1	1	1
P_2	0	—[b]	0.67	1	2	3
P_3	1	0	3.60	1	3	6
P_4	2	2.00	10.29	1	4	10
P_5	3	7.51	22.22	1	5	15
P_6	4	12.00	40.91	1	6	21
P_7	5	19.22	67.85	1	7	28
P_8	6	25.02	104.53	1	8	36
C_1	0	—[b]	0	1	1	1
C_2	1	0	2.00	2	2	4
C_3	3	4.75	9.00	3	3	9
C_4	4	8.00	24.00	4	4	16
C_5	5	11.61	50.00	5	5	25
C_6	6	15.51	90.00	6	6	36
C_7	7	19.65	147.00	7	7	49
C_8	8	24.00	224.00	8	8	64
$K_{1,0}$	0	—[b]	0	1	1	1
$K_{1,1}$	0	—[b]	0.67	1	2	3
$K_{1,2}$	1	0	3.60	1	3	6
$K_{1,3}$	3	4.75	10.29	1	4	11
$K_{1,4}$	6	15.51	22.22	1	5	20
$K_{1,5}$	10	33.22	40.91	1	6	37
K_1	0	—[b]	0	1	1	1
K_2	0	—[b]	0.67	1	2	3
K_3	3	4.75	9.00	3	3	9
K_4	12	43.02	72.00	16	5	38

[a]N.B.: $P_1 \cong K_{1,0} \cong K_1$ = methane, $P_2 \cong K_{1,1} \cong K_2$ = ethane, $P_3 \cong K_{1,2}$ = propane, P_4 = butane, P_5 = pentane, P_6 = hexane, P_7 = heptane, P_8 = octane, C_1 = methylene (carbene), C_2 = ethylene, $C_3 \cong K_3$ = cyclopropane, C_4 = cyclobutane, C_5 = cyclopentane, C_6 = cyclohexane, C_7 = cycloheptane, C_8 = cyclooctane, $K_{1,3}$ = 2–methylpropane (isobutane), $K_{1,4}$ = 2,2–dimethylpropane, $K_{1,5}$ = pentamethylmethane (hypothetical), K_4 = tetrahedrane.
[b]Not defined.

(isobutane)? At best Minoli's criterion 4 (intuitive "feel") is of limited usefulness; at worst it is misleading (vide infra).

$$\chi(G) = [ne/(n+e)]\Sigma_{i<j}\sigma_{ij} \tag{3.1}$$

Bonchev and Polansky [10] listed nine requirements for a complexity index:

(1) It should be independent of the nature of the system.
(2) It should be specified within a unique theoretical conception.
(3) It should take into account the different complexity levels and their hierarchy.
(4) It should exhibit a stronger dependence on the relations between the elements of a system than on the element number.
(5) It should increase monotonically with the number of different complexity features.
(6) It should agree with the intuitive idea of complexity.
(7) It should differentiate the non-isomorphic systems.
(8) It should not be too sophisticated.
(9) It should be applicable to practical purposes.

This list contains several useful generalizations, especially numbers 1, 3 and 5. They were also contained explicitly or implicitly in our early papers on molecular complexity [4–8]. Number 2 is an aesthetic consideration, and number 4 may be true for some systems, but not for others. Number 8 is a subjective matter, which depends upon the mathematical sophistication of the individual. (Ironically, one of the definitions [1] of *sophisticated* is "highly complicated or developed: COMPLEX.") Number 7 may be desirable in some systems, but no quantity is known that fulfills this wish in every case. Number 9 is our raison d'être. The only point we categorically reject is number 6—intuition, which will be discussed below. The solution to our concerns about numbers 2, 4 and 8 is to devise a family of complexity indices of increasing sophistication (i.e., complexity) that can be matched to the requirements of the problem at hand (section 3.2.7).

Branching is an important component of complexity, and Bonchev and Trinajstić [11] proposed specific branching rules: "It is important to notice that our *intuition* provides many more details of branching than is usually supposed. The factors important in branching are (a) the total number of branches, (b) the length of the chain and branches, (c) the position of the branches, and (d) the number of branches attached to a given vertex. ... It is in accordance with our *intuition* to judge that in general the branching should increase when the length of branches increases. However, the branching should decrease relatively when the chain length increases. Examining two other cases, one could conclude that the branching increases with the increased number of branches connected to a given vertex. The second case, in fact, is not so obvious, but it could be associated with the tendency of forming a star-like graph which is more branched than the other ones. The influence of the last factor, the position of branches, could be explained by the combined use of two other factors—the number of branches and their length.... In such a way, one can suppose that branching increases when the branch moves from the end to the middle of the chain. All the examples considered above display the ability of our *intuition* to indicate the general tendency of branching to increase with the total number of branches, the length of branches, their more central position, and their number to a given vertex, and to decrease with the chain length increase. One could try to generalize this tendency in some simple rules on the basis of the quantities connected with the distance matrix (both information measures on graph distances, Wiener number, and the largest eigenvalue of the characteristic polynomial)." Applying such a set of rules in a consistent manner is itself a complicated problem.

One cannot help but be struck by the dependence of all these approaches on 'intuition.' The problem with intuition is that it is not a physical quantity that can be measured, and therefore it is not reproducible and dependable. Moreover, when we are correct in a surmise or 'educated' guess, we credit intuition, but when we are wrong, we don't blame intuition. Finally, one person's intuition is another person's prejudice. An exemplar is the common belief that evolution inexorably results in increasingly complex forms of life, culminating in *Homo sapiens*—us! In recent years this view has been vigorously debated [12–14]. For example, McShea [12] asks, "Has there been a trend in metazoan complexity?" He concludes that it has not been proven, and he observes, "On this question, intuition is probably a poor guide." As C.K. Yoon [15] notes, "One of the reasons that so little has been learned of the evolution of complexity in the century since Charles Darwin first began to ponder the phenomenon in his personal notebooks, is the difficulty of quantifying the complexity that we perceive."

We set out to create a mathematical theory of complexity that would not depend on Minoli's property 4 or number 6 of Bonchev and Polansky, 'intuition.' For example, an axiom-based approach with derived theorems was used to define branching [8]:

Axiom 3.1 *Of two structures resulting from the addition of an atom or group of atoms to a given structure, the one in which it has been affixed to the precursor atom with the higher degree is more branched. (In the context of hydrogen-suppressed graphs, the degree is the number of bonds to other atoms, not including those to hydrogen, see section 3.2.1.)*

Axiom 3.2 *Pairs of structures are ordered by comparing their iterated line graph sequences, so that the one that ultimately dominates corresponds to the more branched structure. (Cf. section 3.2.1 for definitions.)*

Theorem 3.1 *When an additional atom or group of atoms is affixed to a molecule, the resulting structure is more highly branched.*

Theorem 3.2 *The straight-chain molecule (path P_n) is the least-branched member of the family of structures with n atoms and n – 1 bonds.*

Theorem 3.3 *The molecule in which one atom is bonded to n – 1 other atoms (star $K_{1,n-1}$) is the most highly branched structure with n atoms and n – 1 bonds.*

Theorem 3.4 *The branching index BI_M of a graph M is equal to the number of pairs of adjacent lines in M, provided that for a point $BI_M = 0$.*

A branching index BI_M consistent with Axiom 3.1 is given by equation 3.2. Thus, BI_M is the value of the index for molecule M, which is the product of a bond-forming reaction, K + L → M, between molecules K and L with branching indices BI_K and BI_L, respectively; the new bond is formed between an atom of K with degree k and an atom of L with degree l. (Degree is defined in Axiom 1, vide supra.) The number of pairs of adjacent lines is called the number of connections η for short, and its relationship to the degrees d_i ($i = 1, \ldots, n$) of the points in molecular graph M is given by equation 3.3, which is based on the formula for d things taken 2 at a time. The second axiom is necessary because there are pairs of isomers (e.g., 2-methylpentane and 3-methylpentane) that are not ordered by the first. It is a natural extension, as can be appreciated by noting that the pairs of adjacent lines in graph M are the lines in $L(M)$, the line graph of M (section 3.2.1).

$$BI_M = BI_K + BI_L + k + l \tag{3.2}$$

$$BI_M = \eta = \Sigma_i d_i (d_i - 1)/2 \tag{3.3}$$

As far as applicability is concerned, a mathematical model is useful when it clarifies our thinking and even more valuable when it allows us to calculate theoretical values to compare with experimental ones. If the calculation is done before the experiment, then it is called a prediction, which is the goal of any subject that aspires to be a science. If there is a science of complexity [16], this is the standard it must meet.

A number of books have been written about complexity [16–25]. We will discuss some of the concepts from these and other sources, e.g., the roles of information [26–28] and symmetry [29–31], as we develop our theories of molecular complexity and synthetic complexity. In particular, we clarify the oft-quoted generalization that symmetry decreases complexity by showing that it is a simplifying factor for some kinds of complexity, but not for others.

3.2 Can we measure complexity?

Lord Kelvin observed [32], “When you can measure what you are speaking about, and express it in numbers, you know something about it; but when you cannot measure it, when you cannot express it in numbers, your knowledge is of a meager and unsatisfactory kind: it may be the beginning of knowledge, but you have scarcely, in your thoughts, advanced to the stage of *science*, whatever the matter may be.”

Our basic approach to complexity is to abstract a system as a graph and then use the tools of graph theory [33–35] and combinatorics [35, 36] to understand the relationships in it. *System* and *object* are used with their usual definitions [1]—and with the connotation that the former is grander than the latter. For example, a supramolecular complex (system) is made up of molecules (objects), which can be considered the parts of the system that interact and define its complexity. At the next level of “magnification,” the molecule becomes the system, which is made of objects called atoms. Each (neutral) atom is a system comprising a nucleus and one or more electrons. The nucleus is a system of protons and neutrons (except for that of ^{1}H, which has a single proton). Protons and neutrons are systems of fundamental particles, quarks. (Electrons are also fundamental.) Nobody describes supramolecular complexes in terms of quarks and electrons; they are described in terms of molecules or sometimes atoms, e.g., where X-ray crystallographic data are involved. There is a natural hierarchy which enables very complex phenomena to be understood within a conceptually (i.e., mathematically) manageable framework.

3.2.1 Definitions

The definitions given here follow standard texts [33–36]. A *graph G* consists of a finite set $V(G)$ of *vertices* (or *points*) together with a set $E(G)$ of *edges* (or *lines*), which are unordered pairs of distinct vertices of $V(G)$. The *null graph* G_0 has the null set ϕ for $V(G)$ and $E(G)$. An edge $x = uv = vu$ in *G joins* vertices u and v, which are *adjacent vertices*. Vertex u and edge x are *incident* to each other. The *degree d* of a vertex is the number of edges incident to it. Two edges that share a vertex are *adjacent edges*, e.g., uv and vw. Vertex and point are used interchangeably and also edge and line, but vertex is used exclusively with edge and point with line.

When the edges of a graph are directed *arcs*, i.e., ordered pairs of distinct vertices, it is a *directed graph* or *digraph*. In a *multigraph* more than one edge (i.e., a *multiple edge*) joins at least one pair of vertices. A *loop* $x = uu$ is an edge that joins a vertex to itself. A *pseudograph* allows both multiple edges and loops. A graph without multiple edges or loops is referred to as a *normal graph* by some authors [9].

A *path* P_n is a sequence of vertices $v_1, v_2, v_3, \ldots, v_n$ that are joined by edges $v_1v_2, v_2v_3, \ldots, v_{n-1}v_n$. In a *proper* (or *self-avoiding*) *path* each vertex is distinct. (Hence each edge is also distinct.) *Line-disjoint paths* can share points, but not lines. In a *connected graph* all pairs of points are the endpoints of some path. The length of the longest proper path is the *diameter D* of the graph. A *cycle* (or *ring*) C_n is a sequence of vertices $v_1, v_2, v_3, \ldots, v_n$ such that the first vertex v_1 and last one v_n are joined by edge v_1v_n, and all n vertices are distinct. A *tree* is a connected graph without cycles (including loops and multiple edges).

In a *labeled graph* its n points are distinguished from each other by unique labels $l_1, l_2, \ldots, l_n$, e.g., used for testing isomorphism (vide infra). Alternatively, more than one point can have the same label. To avoid confusion these non-unique labels are called *colors*, and to assign them is *to color* the points. A *coloring* of a graph assigns colors to its points in such a way that no two adjacent points have the same color. The colors chosen do not have to be real ones, e.g., an asterisk is sometimes used to represent one color and its absence another. For two colors open and closed circles are usually used. In our usage, a *vertex-labeled (colored) graph* has at least one labeled (colored) vertex that is different from the rest.

A *bipartite graph* or *bigraph* is a graph that has a 2-coloring, i.e., every line joins a point of the first color to a point of the second. All trees are bigraphs, including stars $K_{1,n-1}$ on n points, where one point of the first color is joined to $n - 1$ points of the second.

Two graphs G and H are *isomorphic*, $G \cong H$, if and only if there exists a one-to-one correspondence between their point sets that preserves

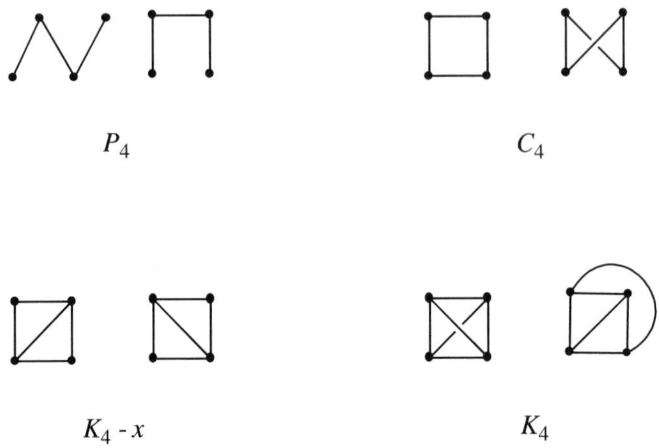

Figure 3.1: Graphs on four vertices that illustrate cyclicity. Two isomorphic presentations are given for each. The corresponding molecules are butane (P_4), cyclobutane (C_4), bicyclobutane (K_{4-x}) and tetrahedrane (K_4).

adjacency. Examples of isomorphic graphs are shown in Figure 3.1. An *invariant* of G is a number $I(G)$ associated with graph G that has the same value for any graph H isomorphic to G. Thus, it does not change when the graph is relabeled, redrawn or manipulated in any way that leaves the adjacency relation intact. For example, the number of points, the number of lines and the number of pairs of adjacent lines are graph invariants. A *topological index* is a real number that is a graph invariant or is derived from one or more invariants.

A *subgraph* S of graph G is a graph that has all its vertices in $V(G)$ and edges in $E(G)$. We include G itself in the set of all possible subgraphs. A *spanning subgraph* of G is a subgraph containing all the vertices of G. A spanning subgraph that is also a tree is a *spanning tree*. The removal of an edge x from G yields the spanning subgraph $G - x$ containing all edges of G except x. For any subset S of $V(G)$, the *subgraph of G induced by S* or *induced subgraph* $\langle S \rangle$ is the maximal subgraph of G with vertex set S, i.e., two vertices are adjacent in $\langle S \rangle$ if and only if they are adjacent in G.

In a *complete graph* K_n each point is joined to every other point, i.e., all pairs of its n points are adjacent. A *clique* in G is a complete subgraph of G. Examples of common cliques in molecular graphs are K_1 (methane), K_2 (ethane), $K_3 \cong C_3$ (cyclopropane) and K_4 (tetrahedrane). The last one is very rare in organic molecules, but not in inorganic and organometallic materials. A *complete bipartite graph* or *complete bigraph* $K_{m,n}$ is a bigraph that contains *m* points of one color, *n* points of a second and all possible *mn* lines. A *biclique* in *G* is a complete bipartite subgraph of *G*. Examples of common bicliques in molecular graphs include $K_{2,2} \cong C_4$ (cyclobutane) and the star graphs $K_{1,n}$ up to $n = 6$ (e.g., homoleptic octahedral transition metal complexes).

If G_1 is a subgraph of G_2, then G_2 is a *supergraph* of G_1. If *u* is a vertex in G_1 and *v* is not, then the addition of edge $x = uv$ to G_1 results in the smallest supergraph $G_2 \cong G_1 + x$ that contains the edge *x*. A chemically relevant example is $G_1 \cong C_3$ (cyclopropane) and $G_2 \cong C_3 + x$ (methylcyclopropane). A *homologous series* $\{H\}$ is a recursively generated sequence of connected graphs G_1, G_2, G_3, …, where G_{i+1} is obtained by incrementing G_i according to a *recurrent rule*, a set of instructions that is repeated at each stage. Examples of homologous series are paths $\{P_n\}$, cycles $\{C_n\}$, star graphs $\{K_{1,n-1}\}$ and complete graphs $\{K_n\}$, which are illustrated in Figure 3.2. (The star graphs can also be symbolized by $\{K_{1,n}\}$, which have $n + 1$ points.)

Molecules are *isomers* whenever they have the same number and kinds of atoms (i.e., the same molecular formula) and the same number of bonds. By analogy, two vertex-colored graphs are *isomers* (or *isomeric graphs*) whenever they have the same number of vertices of each color and the same number of edges, e.g., P_4 (butane) and $K_{1,3}$ (isobutane).

The *line graph* $L(G)$ of a graph *G* has the lines of *G* as its points, and two points in $L(G)$ are adjacent whenever the corresponding lines in *G* are adjacent. Thus, a pair of adjacent lines in *G*, called a *connection*, becomes a line in $L(G)$. The number of connections is symbolized by η, and when counting them in pseudographs, a loop is not adjacent to itself and a double line has one adjacency. Repeating the process embodied in the definition leads to the *iterated line graphs* $L^i(G)$ ($i = 0, 1, 2, \ldots$), where $L^0(G) \cong G$ and $L^1(G) \cong L(G)$. (For completeness, $L(C_1) \cong P_1$, $L(P_1) \cong P_0 \cong G_0$ and $L(G_0)$ is not defined.) The *iterated line graph sequence* is the series of integers N^i, which are the numbers of lines in the iterated line graphs.

An *edge cover C* of graph *G* is any family $C = \{S_1, \ldots, S_n\}$ of subgraphs S_i of *G* with edge sets $E(S_i)$ such that every edge of *G* is in at least one of the $E(S_i)$. In an *edge clique cover* every subgraph $S_i \in C$ is a clique in *G*; in an *edge biclique cover* every $S_i \in C$ is a biclique in *G*. A *minimal edge cover* does not properly contain any other edge cover. Thus, in a minimal edge cover each $S_i \in C$ is essential, i.e., S_i contains at least one edge of *G* that is not contained in any other subgraph $S_j \in C$ $(i \neq j)$. As a test, the

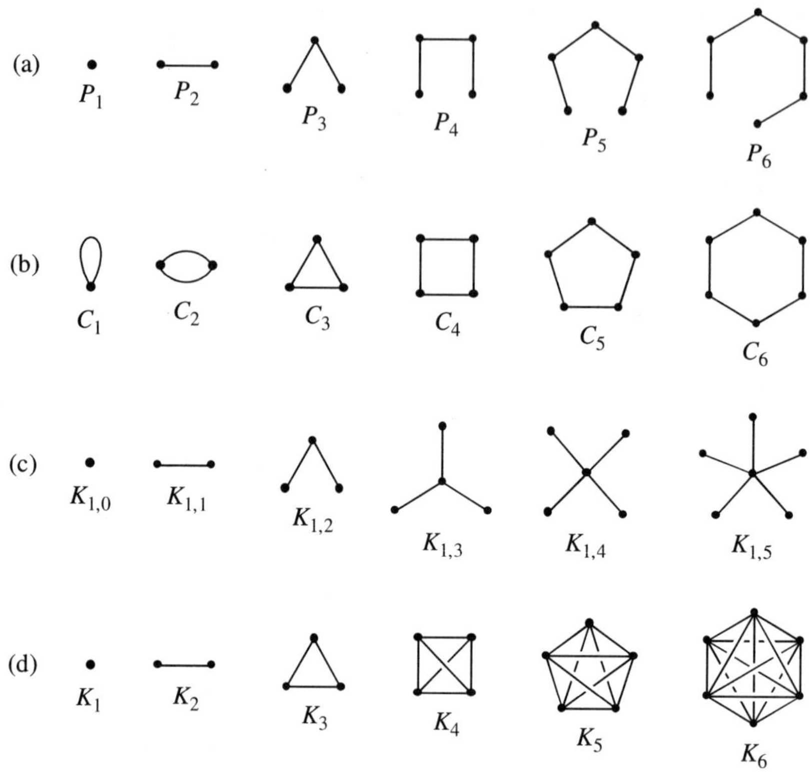

Figure 3.2: The first six graphs in the homologous series (a) P_n, (b) C_n, (c) $K_{1,n-1}$ and (d) K_n for n = 1–6.

exclusion of one S_i at a time results at each turn in a family of subgraphs that no longer covers $E(G)$. A *partition P* is an edge cover with the additional property that each edge belongs to exactly one $S_i \in C$. Thus, in a partition no two subgraphs S_i and S_j $(i \neq j)$ have an edge in common. Figure 3.3 shows the three kinds of minimal edge biclique covers of P_4, including two kinds of partitions.

A molecule can be abstracted as a *molecular graph M* by representing its atoms as vertices and the covalent bonds between them as edges. Ionic interactions are not considered here, but they could be included by adding edges based on distances obtained from X-ray structures or NMR data. Chemical graph theory ordinarily uses *hydrogen-suppressed graphs*, which do not include any hydrogen atoms or the bonds to them. Multiple bonds are represented by multiple edges and lone (unshared) pairs of electrons

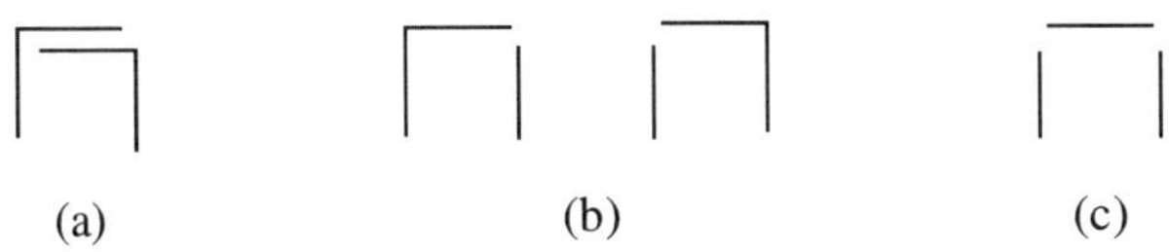

Figure 3.3: Minimal edge biclique covers of P_4 (butane): (a) $2K_{1,2}$ (two propanes), (b) $K_{1,2} + K_{1,1}$ (propane plus ethane) and (c) $3K_{1,1}$ (three ethanes). Covers (b) and (c) are also partitions.

by loops. Heteroatoms can be included by coloring the vertices, where each chemical element is assigned a unique color. Labels on the vertices or edges can be used to convey a wide range of information, such as stereochemistry (*R* or *S* absolute configuration, *E* or *Z* olefin geometry, α or β stereoisomerism), isotopic composition or bond orders.

A synthesis plan can be represented by a *synthesis digraph*, in which the points stand for molecules and the arcs for reactions [37–39]. Since the arcs always point in the forward (synthetic) direction, the arrowheads are omitted in a *synthesis graph*, which is important for its symmetry properties (cf. reflexivity, section 3.3.5) [38–40].

A number of these definitions are illustrated in Figure 3.4, which shows the key structural feature of the β-lactam antibiotics (penicillins, cephalosporins, monobactams). The hydrogen-suppressed molecular graph M_1 contains the loop v_1v_1, which represents the lone pair of electrons on nitrogen. In an ordinary, unstrained amide such as M_3 this pair of electrons is conjugated with the carbonyl group (v_4v_5), and the isomer M_4 is a contributor to the bonding picture via resonance.

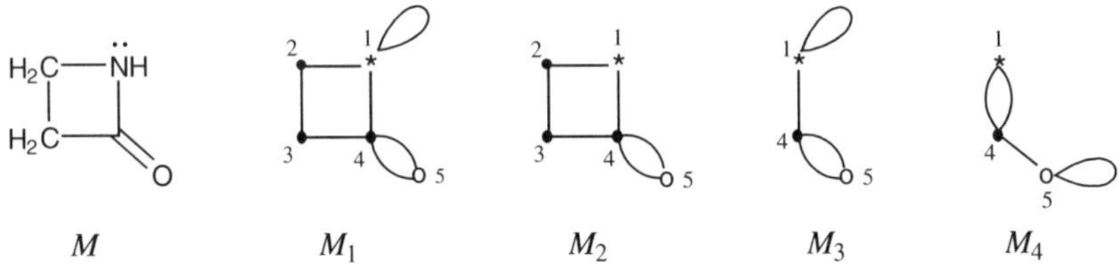

Figure 3.4: The molecule *M* (β-lactam) can be represented by the hydrogen-suppressed molecular graph M_1, which is a pseudograph with loop v_1v_1 and double edge v_4v_5. It is labeled with numbers 1–5 and with three 'colors,' • for C, ∘ for O and $*$ for N. Multigraph M_2 is a spanning subgraph of M_1. For the set of vertices $S = \{v_1, v_4, v_5\}$, M_3 is the induced subgraph $\langle S \rangle$ of M_1 and represents the amide functional group. Graph M_4 is an isomer of M_3.

3.2.2 Kinds of complexity

The two main approaches to measuring the complexity of a system are (i) to examine the steps by which it is constructed, and (ii) to study the way it operates. We call the complexity of construction *extrinsic complexity* and the complexity of operation *intrinsic complexity* [41]. By construction we mean assembly from parts, and by operation we mean the way the parts interact with each other, which defines how they are interrelated. These two kinds of complexity are not independent, e.g., the way the atoms in a molecule interact is determined by their geometric relationships, which also influence how they are assembled when the molecule is synthesized. Most significantly, they differ in their response to symmetry (next section), which can also be used to define them.

The salient feature of extrinsic complexity is the fact that it is not unique. We often do not know how a system was constructed, and more than one way of fabricating it may be possible. (This is true for molecules and is what makes chemical synthesis a fascinating and challenging subject, cf. section 3.3.) For a system that we have not seen constructed, the best we can do is some sort of reverse engineering. Since it is not an intrinsic property of a system, we call this kind of complexity extrinsic.

A variation on the construction theme is to calculate the minimum information required for instructions on how to assemble the system [17–20], which can be thought of as constructing it conceptually. An example of conceptual construction is the "minimum part and placement" description, which has been applied to systems as complex as the biological cell [28]. The complexity of conceptual construction is not an intrinsic property, either, as different instruction sets are usually possible. Furthermore, it is difficult to prove that one of them is minimal. Thus, we can consider the complexity of constructing the physical model or the corresponding conceptual model [42], and we classify both of these variations under extrinsic complexity. We call the complexity of actually fabricating an object the *constructive complexity*. When the object is a molecule or supramolecular complex, it is the *synthetic complexity* (section 3.3.3). For the complexity of conceptual construction, we propose *instructive complexity*.

To measure intrinsic complexity we abstract a system as a graph and then characterize the interactions or other relationships among its parts by using graph invariants. Such relationships are unique for any given system, hence we call this kind of complexity intrinsic. The interactions in the physical model and those in the conceptual model are both subsumed under intrinsic complexity. As a practical matter, we do not distinguish them, since a conceptual model is created when we abstract the system as a graph in the first step of our method. It is also possible to use the minimum information necessary to characterize the interactions in a system as a measure of its complexity [17–20]; however, this approach suffers from the same drawbacks as instructive complexity (vide supra). Conformational complexity (section 3.2.4) and topological complexity (section 3.3.4) are intrinsic.

3.2.3 Symmetry and complexity

It is often said that symmetry decreases complexity, without specifying what kind (cf. section 3.1, Definitions 3, 4A and 4B). Symmetry decreases extrinsic complexity because it reduces the information necessary for instructions on how to assemble the system (instructive complexity) and the work required to actually fabricate it (constructive complexity). On the other hand, the total number of interactions in a system is not decreased when some of them happen to be the same; therefore, intrinsic complexity is not affected by symmetry. (There may be exotic effects, e.g., in spectroscopy some transitions are forbidden by symmetry.) The number of kinds of interactions tends to be decreased by symmetry (cf. diversity, section 3.2.7). The kinds of interactions present in a system are not necessarily unique, and therefore they are not considered intrinsic. For example, many molecules have C-C, C-O and C-N bonds, which are kinds of interactions between atoms; it is the number of each and their arrangement that makes a molecule unique. Thus, sensitivity or insensitivity to symmetry provides another way to define extrinsic and intrinsic complexity, respectively. Rücker and Rücker [43] have discussed *labyrinthicity*, which is independent of symmetry and is therefore a kind of intrinsic complexity.

In the extrinsic view a highly symmetrical graph such as K_n is simple [17], since it is possible to describe one of its points as having degree $d = n - 1$ and then to duplicate it n times. Cycle C_n ($d = 2$) is equally simple in this scheme. In fact, the set of n isolated points $n \times K_1$ ($d = 0$) has essentially the same complexity as K_n. Paths P_n and stars $K_{1,n-1}$ also have significant symmetry and relatively simple descriptions. (Vide infra section 3.2.5 for the recurrent rules that generate these series.)

An example of the effect of symmetry on constructive (synthetic) complexity is the reflexive synthesis of a molecule from two or more identical pieces [38–40], which is more efficient than an alternative synthesis from different ones (section 3.3.5). The presence of symmetry in a molecular graph may or may not result in a more efficient synthesis, depending on whether synthetic methodology exists to take advantage of it [31, 40]. In general, constructive complexity depends on the tools available, in this case synthetic reactions, and it changes with new developments. The intrinsic complexity of a molecule does not change with the state of the art of synthetic chemistry for the simple reason that the structure of the molecule does not change. For this reason we use intrinsic complexity indices to measure molecular complexity.

The "first general index of molecular complexity" was $C_t = C(\eta) + C(E)$, calculated from equation 3.4, where η_i is the number of connections in the ith equivalence class, η is the total number of connections, E_j is the number of atoms of element j and E is the total number of atoms [4]. Thus, an intrinsic complexity index η is used in conjunction with information theory [26, 27] to create a "hybrid" index that takes account of symmetry in addition to other factors. While the function $C(\eta)$ is not an intrinsic complexity index (section 3.2.6), it gives good results in many applications [4, 7, 38]. It can be thought of as the synthetic complexity at a state of the art where any reaction that might be needed to utilize molecular symmetry is known [31].

$$C(\eta) = 2\eta \log_2 \eta - \Sigma_i \eta_i \log_2 \eta_i \tag{3.4a}$$

$$C(E) = E \log_2 E - \Sigma_j E_j \log_2 E_j \tag{3.4b}$$

3.2.4 Complexity factors

Our approach abstracts the system to be studied as a graph G and then uses graph theoretical invariants $I(G)$ as indices of the complexity of G and the system it represents. At the very least, a complexity index must be a monotonically increasing function of those factors that contribute to the complexity of a graph [4–8], viz. size (number of vertices n), branching (degrees d_i of the vertices), edges (e), multiple edges, cycles, loops and labels. *Cyclicity* is defined as the number of cycles divided by the number of vertices (i.e., rings per point) [43]. The graphs in Figure 3.1 illustrate increasing cyclicity, since the number of rings increases while the number of points remains constant. Considering all the relevant factors (lines, cycles, branching), the order of complexity is $P_4 < C_4 < K_4 - x < K_4$, since each successive graph has more of every factor (cf. section 3.2.8). Symmetry is a factor for extrinsic complexity, but not for intrinsic complexity (previous section).

Some of these complexity factors can be studied in homologous series (next section), e.g., path size in $\{P_n\}$, cycle size in $\{C_n\}$, branching at a point in $\{K_{1,n}\}$ and cyclicity in $\{K_n\}$ (cf. section 3.2.1). The number of lines necessarily increases in homologous series of connected graphs. Branching depends on the degrees of the points, and it is often associated with the highest degree of a graph. In a family of isomers on n points, the sum of degrees $\Sigma\, d_i$ ($i = 1, \ldots, n$) is constant, e.g., $\Sigma\, d_i = 6$ for both P_4 and $K_{1,3}$. The latter is more branched, since its highest degree is 3 versus 2 for the former. How many points of degree $d - 1$ in one graph are needed to outweigh one point of degree d in another depends on the mathematical formalism adopted. We will not perseverate on such cases here, as they have been discussed elsewhere [8]. Cyclicity is impossible to separate from branching, since creating a new cycle by joining two points with a new line also increases the degrees of both points. Thus, branching and cyclicity increase in $\{K_n\}$.

A straightforward approach to graph complexity is to take a linear combination of complexity factors, e.g., $C = rR + sS + tT + uU$, where R is the number of rings, S is a size parameter (e.g., the number of vertices), T is the number of trees (a branching index) and U is the number of unsaturations (double edges). The problem with this approach is the choice of coefficients r, s, t and u used to weight the factors, as alluded to by Whitlock [44], who introduced such a scheme for synthetic analysis (cf. section 3.3.2). (Our S should not be confused with his.) Gordon et al. [45] developed the linear combination of graph invariants (LCGI) approach as part of their theory of molecular physical properties.

As Whitlock [44] recognized, his simple LCGI scheme does not incorporate long-range interactions. This deficit can be remedied by including the proper parameters, e.g., our equation for C in the previous paragraph includes, within T, the number of spanning trees, which are global in nature. Lehn [46] has explained why the global view is important: "Complexity implies and results from multiple components and interactions between them with integration, i.e., long-range correlation, coupling and feedback. It is interaction between components that makes the whole more than the sum of the parts and leads to collective properties. Thus, the complexity of an organised system involves three basic features: Complexity = (Multiplicity) (Interaction) (Integration) = MI_2."

Our approach to molecular complexity does not explicitly include conformations, i.e., conformational complexity [47], which may be a drawback in some cases, but is advantageous in others. For example, when searching for reflexive routes to a target molecule (section 3.3.5), our strategy is to examine the molecular graph for isomorphic subgraphs. If conformational information were to be included, many of these would be missed. For example, chair and boat conformations of cyclohexane are different in three dimensions, but they become equivalent when abstracted as graphs.

3.2.5 Homologous series

Homologous series are the scaffolding upon which we build our theory of the complexity of graphs and the molecules represented by them [7, 39, 41, 48]. The first six members of the homologous series discussed here are pictured in Figure 3.2. Starting from P_1, $\{P_n\}$ is the homologous series derived by joining a new vertex v to the vertex u of lowest degree in P_i with a new edge $x = uv$ to obtain P_{i+1}. Upon joining v to P_1, we obtain P_2, which has two vertices of degree 1. Either may be considered the vertex of lowest degree for the next application of the recurrent rule, which gives P_3. In each subsequent step there are two equivalent endpoints of degree 1 from which to choose.

The star graphs $\{K_{1,n}\}$ can be generated by starting with $K_{1,0} \cong P_1$ and joining a new vertex v to the vertex u of highest degree in $K_{1,i}$ with a new edge $x = uv$ to obtain $K_{1,i+1}$. In $K_{1,1} \cong P_2$ either vertex can be chosen for the next application of the recurrent rule, but thereafter we have one vertex of degree $d \geq 2$ and i vertices of degree 1, and the former must be used for each subsequent iteration. The paths $\{P_n\}$ and stars $\{K_{1,n-1}\}$ represent the least and most branched trees on n vertices, respectively.

A third homologous series is $\{C_n\}$, which can be constructed by joining the endpoints of each path P_n with a line to give the corresponding cycle C_n. The 1-cycle C_1 is defined as a loop connecting a point to itself, the 2-cycle C_2 is the smallest multigraph with a double line, and the rest of the n-cycles are rings with n points.

The above homologous series add one new line at a time; it is also possible to add more than one line simultaneously. Starting from $K_1 \cong P_1$, the homologous series of complete graphs $\{K_n\}$ is generated by connecting a new point to all i points of K_i with i new lines to obtain K_{i+1} at each stage. As far as intrinsic complexity is concerned, K_n is the most complex normal graph on n points (cf. next section).

More elaborate homologous series can be imagined, e.g., "ladder" graphs can be built up one rung at a time, or "dendrimers" can be grown by starting with a star and adding more stars at each stage. The simplest homologous series are studied here, as they most clearly embody the complexity factors.

3.2.6 Hierarchies of homologous series

The concept of hierarchy in molecular complexity was introduced in 1983 [7] and expanded in 1987 [10]. *Hierarchies of homologous series* were invented recently as a criterion for indices of intrinsic complexity [41]. The star $K_{1,n-1}$ and the path P_n on n points ($n > 3$) are isomeric trees, i.e., they have the same number of points, the same number of lines and no cycles. Branching is the only relevant complexity factor (section 3.24), and the highest degree in $K_{1,n-1}$ ($d = n - 1$) increases linearly with the number of points n, whereas the highest degree in P_n is constant ($d = 2$). The second highest degree is 1 in each case, and there are no lower degrees. Therefore, graph $K_{1,n-1}$ is more complex than P_n for $n > 3$, and for the two homologous series we say that $\{K_{1,n-1}\}$ dominates $\{P_n\}$ for $n > 3$, symbolically $\{K_{1,n-1}\} > \{P_n\}$ ($n > 3$). For an index of intrinsic complexity, the order of invariants is $I(K_{1,n-1}) \geq I(P_n)$ for all n. (N.B., $K_{1,0} \cong P_1$, $K_{1,1} \cong P_2$ and $K_{1,2} \cong P_3$.)

For the complete graphs K_n and star graphs $K_{1,n-1}$ on n points, three complexity factors must be considered, viz. lines, cycles and branching. Graph K_n has more lines and (by default) more cycles than $K_{1,n-1}$. The highest degree ($d = n - 1$) is the same for both; nevertheless, it is present n times in the former versus once in the latter. Since all relevant complexity factors are greater in the former, we have $\{K_n\} > \{K_{1,n-1}\}$ ($n \geq 3$) and $I(K_n) \geq I(K_{1,n-1})$. (N.B., $K_1 \cong K_{1,0}$ and $K_2 \cong K_{1,1}$.) Thus, we have established a hierarchy for graphs on n points: $\{K_n\} > \{K_{1,n-1}\} > \{P_n\}$ ($n > 3$), and $I(K_n) \geq I(K_{1,n-1}) \geq I(P_n)$ is required for an invariant $I(G)$ to be an index of intrinsic complexity.

The cycle C_n can be derived from the path P_n by connecting its endpoints with a new line, and the degree of branching increases at the points joined. In C_1 a point is joined to itself in a "1-ring," and from the definition of degree, $d = 1$. In C_2 two points are joined by a double line in a "2-ring," and each point has $d = 2$. Thus, C_n has the same number of points as P_n, but one more line and a new ring. We conclude that C_n is more complex than P_n for all n; therefore, $\{C_n\} > \{P_n\}$ and $I(C_n) \geq I(P_n)$. By an analogous argument, K_n is more complex than C_n for all $n > 3$, since it has $n(n-3)/2$ additional lines and more cycles, and we conclude that $\{K_n\} > \{C_n\}$ ($n > 3$) and $I(K_n) \geq I(C_n)$ ($n \geq 3$). Overall, we have the hierarchy $\{K_n\} > \{C_n\} > \{P_n\}$ ($n > 3$), and $I(K_n) \geq I(C_n) \geq I(P_n)$ ($n \geq 3$) for an intrinsic complexity index.

The extension to multigraphs is straightforward. For example, the unbranched 1-alkenes can be derived from the n-alkanes (abstracted as paths P_n) by adding another edge joining adjacent vertices v_1 and v_2 of the molecular graph, where v_1 is an endpoint. Arguments analogous to the ones above give {1-alkenes} > {n-alkanes} and I(1-alkene) $\geq$ I(n-alkane) for corresponding molecular graphs on n points. A similar case can be made for any other branched or unbranched alkene isomer versus the corresponding alkane.

It should be noted that neither $I(K_{1,n-1}) \geq I(C_n)$ nor $I(C_n) \geq I(K_{1,n-1})$ is a requirement for an intrinsic complexity index, as there is no objective way to order the corresponding graphs. Different mathematical formalisms can give different orders, and the one imposed by a particular approach may well be an artifact.

Since it is decreased by symmetry and increased by branching, an index of extrinsic complexity may or may not follow the above relations based on hierarchies of homologous series, depending upon which factor predominates. This in turn depends on the mathematical functions used, e.g., upon going from P_n to C_n, the index $C(\eta)$ (equation 3.4a) is larger for the latter when $n \leq 7$ and smaller when $n > 7$ (Table 3.1). Therefore, $C(\eta)$ cannot be an index of intrinsic complexity.

3.2.7 Complexity indices

An index of intrinsic complexity, which we use to measure molecular complexity, must fulfill three criteria: (i) it must be independent of symmetry, (ii) it must increase as one goes up each homologous series, and (iii) it must increase as one goes up each hierarchy of homologous series at a constant number of vertices, i.e., with the complexity relations established in the previous section: $I(P_n) \leq I(K_{1,n-1}) \leq I(K_n)$ and $I(P_n) \leq I(C_n) \leq I(K_n)$ ($3 \leq n$). For indices that describe the complexity of graphs, continuously increasing functions are preferred, although monotonically increasing functions are allowed. (N.B., the latter can have regions that are constant, and consequently they are also called monotonically non-decreasing functions.)

The simplest invariant that fulfills criteria (i)–(iii) is the number of pairs of adjacent lines η (equation 3.3) [4, 5], which can also be thought of as the number of subgraphs isomorphic to P_3 or the number of paths of length 2. The only simpler invariants are the number of points n (subgraphs isomorphic to P_1 or paths of length 0) and the number of lines e (subgraphs isomorphic to P_2, when the endpoints are included, or paths of length 1), but they do not obey all the hierarchies of homologous series except trivially. For example, neither of them increases upon going from P_n to $K_{1,n-1}$, whereas the number of pairs of adjacent lines increases from $\eta = n - 2$ for P_n to $\eta = (n-1)(n-2)/2$ for $K_{1,n-1}$ ($n > 3$, cf. equation 3.3). (N.B., points and lines are the primitives of graph theory, cf. section 3.2.1.)

The number of ways to cut propane (P_3, "graph 2") out of a molecular graph M_j is the branching index $N_{2,j}$, introduced by Gordon and Kennedy [45]. We generalized it as the number of connections η, i.e., the number of pairs of adjacent lines in *G*, which equals the number of lines in *L*(*G*). Then, multigraphs and pseudographs can be treated by using indices originally defined for normal graphs, provided that a double line is counted as one pair of adjacent lines, and a loop is not adjacent to itself [5, 41]. (N.B., a triple line has three adjacencies.) Selected values of η are given in Table 3.1.

It is natural to ask, what is the second simplest invariant that can be used to measure complexity *C*? Index graphs P_1, P_2 and P_3 were discussed above, and the trees on four points are the path P_4 and the star $K_{1,3}$. For the sake of generality, we limit index graphs to trees, since a cyclic one (e.g., C_3) would give $C = 0$ for all graphs without cycles (trees). Path P_4 can be used to measure intrinsic complexity, provided that the graphs to be examined have diameter $D \geq 3$. Star graphs $K_{1,n-1}$ all have $D \leq 2$, and consequently they have $C = 0$ with P_4 as the index graph. Star $K_{1,3}$ has the same diameter as $P_3 \cong K_{1,2}$ ($D = 2$), but it has maximum $d = 3$, and all graphs with maximum $d \leq 2$ have $C = 0$ with $K_{1,3}$ as the index graph, viz. all paths P_n and cycles C_n. Since $K_{1,3}$ is more complex than P_4 (previous section), the second simplest topological index is the number of subgraphs of *G* isomorphic to P_4, i.e., the number of ways to cut butane out of a molecular graph. For example, the number of butane subgraphs increases for the graphs in Figure 3.1: P_4 (1) < C_4 (4) < $K_4 - x$ (6) < K_4 (12). This index was introduced by Wiener [49] as the polarity number *p*, the number of pairs of atoms separated by three bonds.

The process of testing subgraphs to determine how many of each kind are present in a graph *G* can be continued until all possible subgraphs of *G* have been enumerated, including those with cycles, multiple edges and loops. They provide two powerful new tools, the number of kinds of subgraphs $N_S(G)$ and the total number of subgraphs $N_T(G)$ for any graph *G*. We formulated the "all possible subgraphs" method for measuring similarity in 1986 [50] and showed that similarity could be used to measure complexity; N_T was introduced as an index of complexity in this paper. Bonchev [51] applied N_T to QSAR and used it as the basis for new topological indices. In 1997, Bertz and Sommer [48] used N_T to determine which retrosynthetic bond disconnections (section 3.3.4) give the greatest simplification, and they proposed N_S as a measure of diversity. Bone and Villar [52] published a similar approach to molecular diversity based on induced subgraphs (section 3.2.1), a subset of all possible subgraphs.

Table 3.2: Selected indices based on cliques, bicliques or all possible subgraphs

G	$C_S(G)$	$C_T(G)$	$B_S(G)$	$B_T(G)$	$N_S(G)$	$N_T(G)$
P_1	1	1	1	1	1	1
P_2	2	3	2	3	2	3
P_3	2	5	3	6	3	6
P_4	2	7	3	9	4	10
P_5	2	9	3	12	5	15
P_6	2	12	3	15	6	21
P_7	2	15	3	18	7	28
P_8	2	18	3	21	8	36
C_1	1	1	1	1	2	2
C_2	2	4	2	4	3	5
C_3	3	7	3	9	4	10
C_4	2	8	4	13	5	17
C_5	2	10	3	15	6	26
C_6	2	12	3	18	7	37
C_7	2	14	3	21	8	50
C_8	2	16	3	24	9	65
$K_{1,0}$	1	1	1	1	1	1
$K_{1,1}$	2	3	2	3	2	3
$K_{1,2}$	2	5	3	6	3	6
$K_{1,3}$	2	7	4	11	4	11
$K_{1,4}$	2	9	5	20	5	20
$K_{1,5}$	2	11	6	37	6	37
K_1	1	1	1	1	1	1
K_2	2	3	2	3	2	3
K_3	3	7	3	9	4	10
K_4	4	15	4	26	10	64

Diversity is a kind of extrinsic complexity, as it is intimately connected to symmetry (section 3.2.3).

The number of kinds of subgraphs $N_S(G)$ counts non-isomorphic, connected subgraphs, and the total number of subgraphs $N_T(G)$ counts all possible connected subgraphs of graph G (Table 3.2). The former is an extrinsic complexity index, as it can be affected by symmetry (section

3.2.9). The latter is an intrinsic complexity index, as it is independent of symmetry and is a continuously increasing function for all four homologous series. Furthermore, it obeys the relations based on hierarchies (for $n \geq 3$ points), e.g., $[N_T(P_4) = 10] \leq [N_T(K_{1,3}) = 11] \leq [N_T(K_4) = 64]$ and $[N_T(P_4) = 10] \leq [N_T(C_4) = 17] \leq [N_T(K_4) = 64]$. Each point P_1 is considered to be a subgraph, and so is G itself. Defined in this way, $N_S(G)$ and $N_T(G)$ can be calculated from concise formulas for specific homologous series: $N_S(P_n) = n$ and $N_T(P_n) = n(n + 1)/2$; $N_S(C_n) = n + 1$ and $N_T(C_n) = n^2 + 1$; $N_S(K_{1,n}) = n + 1$ and $N_T(K_{1,n}) = 2^n + n$. (N.B., the path P_n and the star $K_{1,n-1}$ on n points have the same value of $N_S = n$.) Rücker and Rücker [53] have discovered formulas for $N_S(K_n)$ and $N_T(K_n)$, and they have published a computer program for all possible substructures.

Index N_T is highly discriminating, but it can be difficult to determine for large structures. When one cannot enumerate all possible subgraphs, there are selected sets of them that can be investigated. The total number of trees T_T and the number of kinds of trees T_S (Table 3.1) provide good approximations of N_T and N_S, respectively, for many problems. The number of spanning trees T_{span} was introduced by Tutte [54] to measure the complexity of networks. It appears to work well for graphs with varying degrees of cyclicity [55]; however, all trees have but one spanning subgraph, the tree itself. Two interesting classes of subgraphs are cliques and bicliques, which have a rich mathematical history [33, 34]. The total numbers of cliques C_T and bicliques B_T and the corresponding numbers of kinds C_S and B_S are listed for selected graphs in Table 3.2. The level of sophistication chosen depends on what is needed to solve a problem. There can be complications in $C_T(G)$ and $B_T(G)$ upon labeling G (section 3.2.9).

We recently introduced several new topological indices based on edge clique or biclique covers [41]: the number of kinds of minimal edge clique covers k_S, the total number of minimal edge clique covers k_T, the number of kinds of minimal edge biclique covers k_S^{bi}, and the total number of minimal edge biclique covers k_T^{bi} (cf. section 3.2.1). Figure 3.3 shows the four minimal edge biclique covers for P_4 (butane), so that $k_T^{bi} = 4$. There are three different kinds, $2K_{1,2}$, $K_{1,2} + K_{1,1}$ and $3K_{1,1}$, and $k_S^{bi} = 3$. The last one is also a minimal edge clique cover ($3K_{1,1} \cong 3K_2$), and $k_T = k_S = 1$. The two covers labeled $K_{1,2} + K_{1,1}$ are isomorphic; however, in general two covers do not have to be isomorphic to be of the same kind. They are the same kind of cover whenever they consist of the same subgraphs in equal numbers. In contrast, subgraphs are the same kind if and only if they are isomorphic.

The numbers of kinds of partitions p_S and p_S^{bi} and the total numbers of partitions p_T and p_T^{bi} of the edges of graph G into cliques and bicliques, respectively, are usually easier to determine than the analogous k-indices, especially for large graphs. In Figure 3.3, $K_{1,2} + K_{1,1}$ and $3K_{1,1}$ are edge biclique partitions, and $3K_{1,1} \cong 3K_2$ is an edge clique partition. The new k- and p-indices are compiled in Table 3.3, except for p_S and p_T, which are the same as k_S and k_T for most of the graphs listed. (N.B., they diverge at $n = 4$ for K_n, viz. $p_S(K_4) = 3$ and $p_T(K_4) = 6$.)

The invariants based on clique covers, k_S, k_T, p_S and p_T, are not generally applicable, owing to almost complete degeneracy in $\{P_n\}$, $\{C_n\}$ and $\{K_{1,n-1}\}$. The bipartite indices k_T^{bi} and p_T^{bi} appear to be good intrinsic complexity indices. They are continuously increasing functions for all four homologous series, obey all the complexity relations based on hierarchies and are independent of symmetry. The indices k_S^{bi} and p_S^{bi} are continuously or monotonically increasing functions, respectively, except for the presence of an extra minimal edge biclique cover for $C_4 \cong K_{2,2}$, which is also a partition. They are not intrinsic, since they can be affected by symmetry.

This method for measuring molecular complexity and molecular diversity is very general. Double bonds are represented by multiple edges in multigraphs and lone pairs of electrons (lpe) by loops in pseudographs, and both of these factors increase N_S and N_T relative to the corresponding normal graphs on the same vertices. Heteroatoms can be included by coloring the vertices of the molecular graph, and other labels can be used to further differentiate the vertices or edges, e.g., according to geometric considerations (stereochemistry) or intensities of interactions (bond orders). Such labeling can affect N_S, but not N_T (section 3.2.9). This is the only complexity factor that is not accounted for by N_T. (N.B., symmetry is not a factor for intrinsic complexity, cf. section 3.2.4.) When a heteroatom has a lpe, a loop can be added in addition to the label, and N_T increases. Since most non-metallic elements have at least one lpe, this modification is very useful (section 3.3.4). In reactions that involve *d* or *f* electrons, the occupied orbitals can also be represented by loops.

The all possible subgraphs method is not limited to molecules. The vertices of a graph can represent the parts of any system, and they can be labeled with as many different colors as there are different kinds of parts. McKinney [13] writes in *Science*, "Of course complexity itself can be a fuzzy concept, but Bonner is clear about what he means: increasing complexity is associated not only with increasing number of components but also with increasing number of kinds of components." Our method gives this information in terms of N_T and N_S in a conceptually straightforward way.

Table 3.3: Minimal edge clique and biclique covers and edge biclique partitions[a]

G	$k_S(G)$	$k_T(G)$	$k_S^{bi}(G)$	$k_T^{bi}(G)$	$p_S^{bi}(G)$	$p_T^{bi}(G)$
P_1	0[b]	0[b]	0[b]	0[b]	0[b]	0[b]
P_2	1	1	1	1	1	1
P_3	1	1	2	2	2	2
P_4	1	1	3	4	2	3
P_5	1	1	4	7	3	5
P_6	1	1	5	13	3	8
P_7	1	1	7	24	4	13
P_8	1	1	8	44	4	21
C_1	0[c]	0[c]	0[c]	0[c]	0[c]	0[c]
C_2	1	1	1[c]	1[c]	1[c]	1[c]
C_3	2	2	3	7	2	4
C_4	1	1	5	12	4	8
C_5	1	1	5	21	3	11
C_6	1	1	7	39	4	18
C_7	1	1	8	71	4	29
C_8	1	1	10	131	5	47
$K_{1,0}$	0[b]	0[b]	0[b]	0[b]	0[b]	0[b]
$K_{1,1}$	1	1	1	1	1	1
$K_{1,2}$	1	1	2	2	2	2
$K_{1,3}$	1	1	4	8	3	5
$K_{1,4}$	1	1	9	49	5	15
$K_{1,5}$	1	1	20	522	7	52
K_1	0[b]	0[b]	0[b]	0[b]	0[b]	0[b]
K_2	1	1	1	1	1	1
K_3	2	2	3	7	2	4
K_4	5	16	—[d]	—[d]	7	70

[a]See footnote a, Table 3.1. [b]These values are different from those in [41], which followed [34]; as in [33], we do not consider K_1 to be a clique in edge covers here. [c]These values are different from those in [41] to be consistent with the way the new *B*- and *C*-indices are calculated, i.e., $K_{1,1}$ does not cover C_1 and $K_{1,2}$ does not cover C_2. [d]Not yet calculated.

3.2.8 Sample calculations

To illustrate the calculation of the indices discussed above, we derive them for P_4 (butane), C_4 (cyclobutane), $K_4 - x$ (bicyclobutane) and K_4 (tetrahedrane), the molecular graphs in Figure 3.1. In P_4 the kinds of subgraphs (number of each) are P_1 (4), P_2 (3), P_3 (2) and P_4 (1). Therefore, $N_S = T_S = 4$ and $N_T = T_T = 10$, as there are no cycles. There are two kinds of cliques, K_1 (4) and K_2 (3), and three kinds of bicliques, $K_{1,0}$ (4), $K_{1,1}$ (3) and $K_{1,2}$ (2); therefore, $C_S = 2$, $C_T = 7$, $B_S = 3$ and $B_T = 9$. (N.B., $K_{1,0} \cong K_1 \cong P_1$, $K_{1,1} \cong K_2 \cong P_2$ and $K_{1,2} \cong P_3$, see Figure 3.2.) The only minimal edge clique cover is $3K_2$, which is also a partition, and $k_S = k_T = p_S = p_T = 1$. As shown in Figure 3.3, there are three kinds of minimal edge biclique covers: $2K_{1,2}$ (1), $K_{1,2} + K_{1,1}$ (2) and $3K_{1,1}$ (1), and the last two are partitions; therefore, $k_S^{bi} = 3$, $k_T^{bi} = 4$, $p_S^{bi} = 2$ and $p_T^{bi} = 3$.

In C_4 the kinds of subgraphs (number of each) are P_1 (4), P_2 (4), P_3 (4), P_4 (4) and C_4 (1); therefore, $N_S = 5$ and $N_T = 17$. There is one cyclic subgraph, the biclique $K_{2,2} \cong C_4$ itself, so that $T_S = 4$ and $T_T = 16$. The cliques are K_1 (4) and K_2 (4), and the bicliques are $K_{1,0}$ (4), $K_{1,1}$ (4), $K_{1,2}$ (4) and $K_{2,2}$ (1); therefore, $C_S = 2$, $C_T = 8$, $B_S = 4$ and $B_T = 13$. The only minimal edge clique cover is $4K_2$, which is also a partition; therefore, $k_S = k_T = p_S = p_T = 1$. The minimal edge biclique covers are $K_{2,2}$ (1), $2K_{1,2}$ (2), $2K_{1,2} + K_{1,1}$ (4), $K_{1,2} + 2K_{1,1}$ (4) and $4K_{1,1}$ (1); therefore, $k_S^{bi} = 5$ and $k_T^{bi} = 12$. Except for $2K_{1,2} + K_{1,1}$ (4), they are partitions; consequently, $p_S^{bi} = 4$ and $p_T^{bi} = 8$.

In $K_4 - x$ the kinds of subgraphs (number of each) are P_1 (4), P_2 (5), P_3 (8), C_3 (2), P_4 (6), C_4 (1), $K_{1,3}$ (2), $C_3 + x$ (4) and $K_4 - x$ (1); therefore, $N_S = 9$ and $N_T = 33$. Excluding subgraphs with cycles, $T_S = 5$ and $T_T = 25$. There is a new kind of clique, $K_3 \cong C_3$, and a new kind of biclique, $K_{1,3}$. The cliques are K_1 (4), K_2 (5) and K_3 (2), so that $C_S = 3$ and $C_T = 11$, and the bicliques are $K_{1,0}$ (4), $K_{1,1}$ (5), $K_{1,2}$ (8), $K_{1,3}$ (2) and $K_{2,2}$ (1), so that $B_S = 5$ and $B_T = 20$. The minimal edge clique covers are $2K_3$ (1), $K_3 + 2K_2$ (2) and $5K_2$ (1), and the last two are partitions; therefore, $k_S = 3$, $k_T = 4$, $p_S = 2$ and $p_T = 3$. The minimal edge biclique covers are $K_{2,2} + K_{1,3}$ (2), $K_{2,2} + K_{1,2}$ (4), $K_{2,2} + K_{1,1}$ (*1*), $2K_{1,3}$ (1), $K_{1,3} + 2K_{1,2}$ (8), $K_{1,3} + K_{1,2} + K_{1,1}$ (8), $K_{1,3} + K_{1,2}$ (2), $K_{1,3} + 2K_{1,1}$ (2), $4K_{1,2}$ (1), $3K_{1,2}$ (16), $3K_{1,2} + K_{1,1}$ (8), $2K_{1,2} + 2K_{1,1}$ (18), $2K_{1,2} + K_{1,1}$ (*10*), $K_{1,2} + 3K_{1,1}$ (*8*) and $5K_{1,1}$ (*1*), from which we derive $k_S^{bi} = 15$, $k_T^{bi} = 90$, $p_S^{bi} = 6$ and $p_T^{bi} = 24$. (The numbers of partitions are italicized in this list.)

In K_4 the kinds of subgraphs (number of each) are P_1 (4), P_2 (6), P_3 (12), C_3 (4), P_4 (12), C_4 (3), $K_{1,3}$ (4), $C_3 + x$ (12), $K_4 - x$ (6) and K_4 (1); therefore, $N_S = 10$, $N_T = 64$, $T_S = 5$ and $T_T = 38$. There is one new kind of clique, K_4 itself. Then, the cliques are K_1 (4), K_2 (6), K_3 (4) and K_4 (1), so that $C_S = 4$ and $C_T = 15$; the bicliques are $K_{1,0}$ (4), $K_{1,1}$ (6), $K_{1,2}$ (12), $K_{1,3}$ (4) and $K_{2,2}$ (3), so that $B_S = 5$ and $B_T = 29$. The minimal edge clique covers are K_4 (1), $3K_3$ (4), $2K_3 + K_2$ (6), $K_3 + 3K_2$ (4) and $6K_2$ (1), of which the first and last two are partitions; therefore, $k_S = 5$, $k_T = 16$, $p_S = 3$ and $p_T = 6$. As far as minimal edge biclique covers are concerned, k_S^{bi} and k_T^{bi} are relatively large and have not been enumerated, but the partitions are manageable. Thus, $p_S^{bi} = 7$ and $p_T^{bi} = 70$, which are derived from $K_{2,2} + 2K_{1,1}$ (3), $K_{1,3} + K_{1,2} + K_{1,1}$ (12), $K_{1,3} + 3K_{1,1}$ (4), $3K_{1,2}$ (8), $2K_{1,2} + 2K_{1,1}$ (30), $K_{1,2} + 4K_{1,1}$ (12) and $6K_{1,1}$ (1).

All the invariants discussed above give the same order of complexity for the graphs considered, viz. $I(P_4) \leq I(C_4) \leq I(K_4 - x) \leq I(K_4)$, as summarized in Table 3.4. Some of the invariants are indices of intrinsic complexity, others extrinsic. Each successive graph has more edges, cycles and branching (cf. section 3.2.4) than the one before it, and any decrease in an extrinsic complexity index caused by increased symmetry is outweighed by the other factors.

Table 3.4: Index values for selected graphs[a]

index	P_4	C_4	$K_4 - x$	K_4
N_T	10	17	33	64
N_S	4	5	9	10
T_T	10	16	25	38
T_S	4	4	5	5
B_T	9	13	20	29
B_S	3	4	5	5
C_T	7	8	11	15
C_S	2	2	3	4
k_T	1	1	4	16
k_S	1	1	3	5
p_T	1	1	3	6
p_S	1	1	2	3
k_T^{bi}	4	12	90	—[b]
k_S^{bi}	3	5	15	—[b]
p_T^{bi}	3	8	24	70
p_S^{bi}	2	4	6	7

[a]N.B.: P_4 = butane, C_4 = cyclobutane, $K_4 - x$ = bicyclobutane, K_4 = tetrahedrane. [b]Not yet calculated.

3.2.9 Vertex-labeled graphs

Vertices of graphs can be labeled to convey information about many structural details, e.g., heteroatoms, isotopic composition or chirotopicity. This can affect the various indices in different ways, e.g., coloring a vertex to represent a heteroatom can change the symmetry of a system, and some indices are sensitive to symmetry, whereas others are not. As shown below, adding a second label does not always change the value of an index, even when the first one does.

In the "all possible subgraphs" method (section 3.2.7), the total number of subgraphs $N_T(G)$ does not change upon coloring one or more of the vertices of G, but the number of kinds of subgraphs $N_S(G)$ increases by at least 1 (except for $G \cong P_1$). The choice of colors is arbitrary, so that interchanging two colors does not change the results. A simple example is propane (•–•–•), which has $N_S = 3$ kinds of subgraphs (•–•–•, •–•, •). When one endpoint is colored, there are $N_S = 5$ kinds (○–•–•, ○–•, •–•, ○, •), and when the center point is colored, there are $N_S = 4$ kinds (•–○–•, ○–•, ○, •); nevertheless, in all cases there are $N_T = 6$ total subgraphs. Another way to look at these examples is to say that upon going from ethanol (○–•–•) to dimethyl ether (•–○–•) to propane (•–•–•), symmetry increases, and the number of kinds of subgraphs (i.e., diversity) decreases from $N_S = 5$ to 4 to 3, respectively.

If the center point of each of these graphs is labeled to represent isotopic composition, then for dimethyl ether N_S remains the same, whereas for propane and ethanol N_S increases by 1. Again, N_T does not change. For a complete treatment of vertex-labeled graphs, one must consider both intrinsic and extrinsic complexity, e.g., both N_T and N_S.

On account of the way they are defined, there is a subtle effect of coloring vertices on the numbers of kinds of cliques and bicliques, which can have a dramatic effect on the total numbers of these subgraphs. In an unlabeled graph such as •–•–•, the subgraphs •, •–• and •–•–• are bicliques, since they can be colored as desired for the purpose of classification. However, in a graph with two colors such as •–○–• or ○–•–•, the vertices are already colored. In these graphs the subgraphs •, ○, ○–• and •–○–• are bicliques, but •–• and ○–•–• are not, since from the definition of a bigraph, every vertex of the first color must be connected only to vertices of the second (section 3.2.1). By convention, all the vertices in a clique must have the same color, and in the examples above the cliques are ○, • and •–•. Thus, while N_T and T_T are not changed by labeling, C_T and B_T may be changed by coloring vertices. In those cases where one wishes to avoid this idiosyncrasy, the colors can be ignored.

3.3 Synthesis

Synthesis involves putting together the parts of a system to make the whole. The way they interact is what makes the whole more than the sum of its parts (intrinsic complexity), and the way they are arranged can have a profound effect on the most efficient way to fabricate the system (constructive complexity). The *synthetic complexity* is the constructive complexity of preparing a molecule or supramolecular complex, and there are many factors that contribute to it: the structural details of the target molecule (molecular complexity), the synthetic reactions available (state of the art of synthetic chemistry), methods of purification (state of the art of separations science), and accessibility of starting materials. The availability and cost of starting materials depends on the level of development of industrial chemistry and on natural sources such as petroleum or biomass. Wender has discussed some of these issues in the context of the ideal synthesis [56], and Hudlicky has emphasized the importance of practicality [57].

Once a target is chosen, these factors are outside the immediate control of the synthetic chemist. What he or she can control are the precise starting materials and reactions selected, i.e., the synthetic plan. We focus on the intrinsic complexity of molecules (molecular complexity) and how it changes during synthetic reactions and over the course of multistep syntheses. An academic researcher may have larger latitude in the choice of targets and reactions than an industrial scientist, but many of the challenges they face are the same. The examples below were chosen to illustrate the principles involved and were kept simple enough that the calculations can be performed without a computer. David Bradley observed in the *New Scientist* [58], "It is rather astonishing how little chemists really know about designing the shortest and simplest synthesis of a new molecule."

3.3.1 Synthetic reactions

Only some chemical reactions qualify as general synthetic reactions. They must give high yields with a range of substrates and afford products that people want. If the products are valuable enough, expensive starting materials, reagents or catalysts can be used. Generally speaking, the most valuable synthetic reactions are those that rapidly build up the molecular skeleton and are associated with large increases in molecular complexity. This helps minimize the number of steps required to prepare a target molecule, and each step represents the expenditure of time and money. Each step also exposes the evolving structure to reagents or energy that can induce side reactions. While the yields of synthetic reactions tend to be high, they are rarely 100%, and therefore more steps usually means a lower overall yield, which has been referred to as the "arithmetic fiend" [59]. This argument is similar to those for convergent syntheses (section 3.3.5) and the minimization of excess complexity (section 3.3.2).

New synthetic reactions are being developed every year, and the question may be asked, how do they compare to previous ones in terms of their ability to efficiently construct molecules? One application of an intrinsic complexity index is to measure the power of synthetic reactions to increase molecular complexity upon going from starting materials to products [4, 38]. In this way an index of molecular complexity can serve as a yardstick to gauge the state of the art of synthetic chemistry. Another approach is to calculate the average change in complexity per step in a multistep synthesis (section 3.3.2) and monitor how it evolves over time.

To calculate the change in complexity ΔC for a reaction, the sum of the C-values for the starting materials is subtracted from the sum for the products (equation 3.5a). Until it is established that an invariant I measures complexity, we use ΔI (equation 3.5b). (In the tables we use Δ for convenience.) When two or more identical molecules are joined, the initial complexity is just that of one of them, since there is only one reactant that can undergo yield-decreasing side reactions. Moreover, unless they are starting materials in a route, there is one intermediate that contributes to excess complexity (section 3.3.2). On the other hand, for disconnections all the resulting fragments are included in the calculation, since even when two of them are the same in a disconnection, they need not be identical in the corresponding transform (section 3.3.4).

To calculate the sum of "total" indices $I_T(G_1)$ and $I_T(G_2)$ for graphs G_1 and G_2, one simply adds them (equation 3.6), and to calculate the sum of "kinds of" indices $I_S(G_1)$ and $I_S(G_2)$, one adds them and then subtracts the number of duplicates I_D (equation 3.7). For three or more molecules, it is useful to consider the ensemble of starting materials or products and count connected subgraphs as usual. The same rules apply to covers.

$$\Delta C(\text{reaction}) = \Sigma C(\text{products}) - \Sigma C(\text{starting materials}) \tag{3.5a}$$

$$\Delta I(\text{reaction}) = \Sigma I(\text{products}) - \Sigma I(\text{starting materials}) \tag{3.5b}$$

$$I_T(G_1 + G_2) = I_T(G_1) + I_T(G_2) \tag{3.6}$$

$$I_S(G_1 + G_2) = I_S(G_1) + I_S(G_2) - I_D \tag{3.7}$$

For example, there are $N_S = 7$ kinds of subgraphs for the acetaldehyde graph G_1 (○ = • – •, ○ – • – •, ○ = •, ○ – •, • – •, ○, •) and $N_S = 7$ for the oxygen-labeled acetaldehyde graph G_2 (* = • – •, * = • – •, * = •, * – •, • – •, *, •); both have $N_T = 10$. There are $N_T(G_1 + G_2) = 10 + 10 = 20$ total subgraphs and $N_S(G_1 + G_2) = 7 + 7 - 2 = 12$ kinds of subgraphs for the ensemble $G_1 + G_2$, since there are $I_D = 2$ duplicates (• and • – •). Lone pairs of electrons on heteroatoms are not included, unless they are involved in a reaction (e.g., R_3P: + allyl bromide, section 3.3.2) or needed to differentiate disconnections (section 3.3.4).

Some examples of classic synthetic reactions are the Aldol, Diels-Alder, Wittig and Reppe-Vollhardt reactions, which are abstracted in Figures 3.5–3.8. Other important developments such as the Pauson-Khand and House-Whitesides reactions, catalytic hydrogenation and sigmatropic rearrangement (Figures 3.9–3.12) are also analyzed. More information about these reactions can be found in organic chemistry textbooks [60].

In the typical Aldol reaction a new σ bond is formed between the carbon atom adjacent to a carbonyl group of an aldehyde or ketone and the carbon of a second such carbonyl, which becomes a hydroxyl group. One new carbon-carbon σ bond is formed at the expense of a carbon-oxygen π bond, whether it is the simple, archetypal dimerization of acetaldehyde to aldol (Figure 3.5) or the "complex Aldol reactions" used to synthesize phorboxazole B [61]. Table 3.5 contains the values of a number of complexity indices based on subgraphs for the former. Since the molecular graphs do not have high cyclicities, the total number of subgraphs N_T can be approximated by the total number of trees T_T; thus, $\Delta N_T = 36 - 10 = 26$ and $\Delta T_T = 30 - 8 = 22$. The same is true for kinds of subgraphs; thus, $\Delta N_S = 12$ and $\Delta T_S = 8$. The indices based on cliques or bicliques give little or no change in complexity. (If the colors are ignored, then $\Delta B_T = 11$ and $\Delta C_T = 6$.) The number of heteroatoms does not change during the course

(*) (*)

1 **1 (1*)** **2 (2*)**

Figure 3.5: Aldol reaction of acetaldehyde (**1**) with itself or oxygen-labeled acetaldehyde (**1***), affording aldol (**2**) or labeled aldol (**2***).

Table 3.5: Aldol reaction of acetaldehyde (**1**) with itself or oxygen-labeled acetaldehyde (**1***), affording aldol (**2**) or labeled aldol (**2***)

Index	1 (1*)	1+1 (1+1*)	2 (2*)	Δa(Δb*)
N_T	10 (10)	10 (20)	36 (36)	26 (16)
N_S	7 (7)	7 (12)	19 (23)	12 (11)
T_T	8 (8)	8 (16)	30 (30)	22 (14)
T_S	5 (5)	5 (8)	13 (17)	8 (9)
B_T	5 (5)	5 (10)	9 (9)	4 (–1)
B_S	3 (3)	3 (5)	3 (5)	0 (0)
C_T	4 (4)	4 (8)	9 (9)	5 (1)
C_S	3 (3)	3 (4)	3 (4)	0 (0)

of the Aldol reaction, so that ΔN_T and ΔT_T are good approximations of the change in molecular complexity.

When the starting materials are different, the change in molecular complexity is lower, since the initial complexity is higher. The values for the labeled Aldol reaction (parentheses in Table 3.5) are for the case where one of the oxygen atoms is isotopically labeled (e.g., ^{18}O). Then, for molecular complexity we have $\Delta N_T = 36 - (10 + 10) = 16$ and $\Delta T_T = 30 - (8 + 8) = 14$. The changes in diversity, $\Delta N_S = 11$ and $\Delta T_S = 9$, are essentially the same as for the unlabeled case. The difference between the labeled and unlabeled Aldol reactions is an illustration of reflexivity (section 3.3.5).

The archetypical Diels-Alder reaction is illustrated in Figure 3.6. Two new σ bonds are formed at the net expense of two π bonds with the concommitant formation of a new ring. The increase in molecular complexity, $\Delta N_T = 38$ (Table 3.6), is more than twice that for the labeled Aldol reaction ($\Delta N_T = 16$), whereas the increase in diversity, $\Delta N_S = 9$, is approximately the same ($\Delta N_S = 11$, vide supra). Since this Diels-Alder reaction has two different reactants, we compare it to the labeled Aldol reaction, which also has two starting materials. The approximation based on trees is a good one for the total number, $\Delta T_T = 30$, but not for the number of kinds, $\Delta T_S = 2$. We do not think it is valid to interpret such results too minutely, e.g., we do not attach much meaning to the difference between 38 and 30. On the other hand, we believe the difference between 38 and 16 is significant.

In the Wittig reaction of an aldehyde with a phosphorane (Figure 3.7), the carbon-oxygen and phosphorus-carbon double bonds in the starting materials are exchanged for carbon-carbon and phosphorus-oxygen double

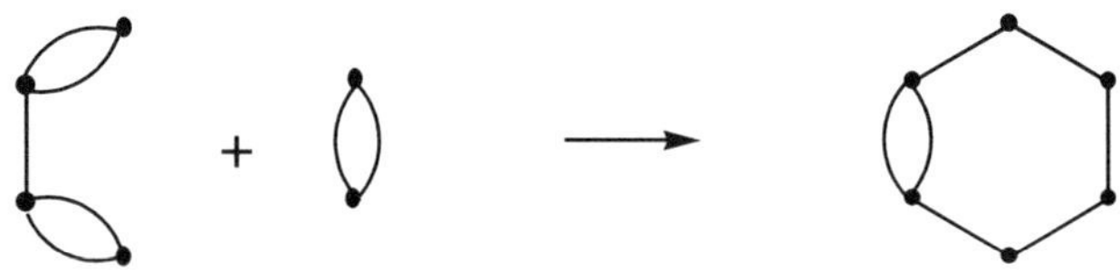

3 4 5

Figure 3.6: Diels-Alder reaction of butadiene (**3**) and ethylene (**4**) to cyclohexene (**5**).

Table 3.6: Diels-Alder reaction of butadiene (**3**) and ethylene (**4**) to cyclohexene (**5**)

Index	3	4	3 + 4	5	Δ
N_T	26	5	31	69	38
N_S	8	3	8	17	9
T_T	17	4	21	51	30
T_S	4	2	4	6	2
B_T	13	4	17	21	4
B_S	3	2	3	3	0
C_T	9	4	13	13	0
C_S	2	2	2	2	0

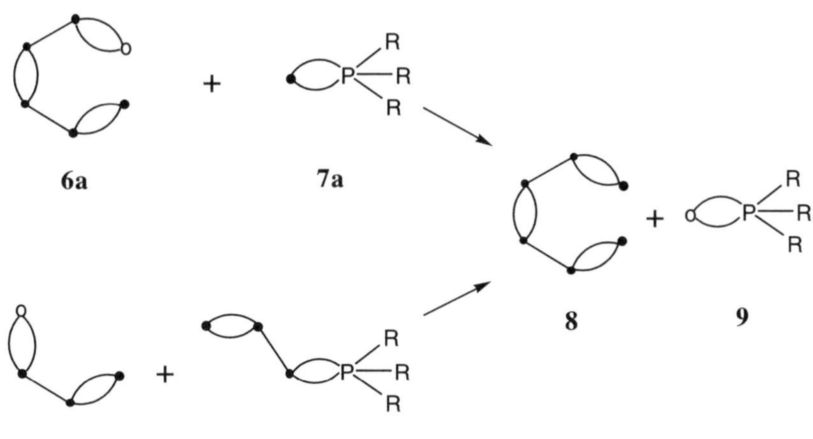

6a 7a 8 9 6b 7b

Figure 3.7: Wittig reactions of aldehydes (**6a**, 2,4-pentadienal; **6b**, acrolein) with phosphoranes (**7a**, methylene; **7b**, allylidene), affording 1,3,5-hexatriene (**8**) and phosphine oxide (**9**).

Table 3.7: Wittig reactions of aldehydes (**6a**, 2,4-pentadienal; **6b**, acrolein) with phosphoranes (**7a**, methylene; **7b**, allylidene), affording 1,3,5-hexatriene (**8**) and phosphine oxide (**9**)

Index	6a (6b)	7a (7b)	8	9	Δa(Δb)
N_T	95 (26)	36 (141)	95	36	0 (–36)
N_S	34 (14)	14 (42)	19	14	–14 (–18)
T_T	48 (17)	28 (83)	48	28	0 (–24)
T_S	11 (7)	10 (20)	6	10	–4 (–8)
B_T	8 (6)	14 (16)	22	14	14 (14)
B_S	3 (3)	7 (7)	3	7	1 (1)
C_T	12 (7)	5 (10)	14	5	2 (2)
C_S	3 (3)	3 (4)	2	3	0 (0)

bonds in the products. The residues (R) on phosphorus are usually phenyl groups, which are inert under the standard reaction conditions. Since this group has a relatively high complexity (vide infra), which does not change during the course of the reaction, it is abstracted as a single vertex ®. Except for one labeled vertex, the molecular graphs for the starting aldehyde and the hydrocarbon product are isomorphic, and the same is true for the starting phosphorane and phosphine oxide by-product; therefore, $\Delta N_T = \Delta T_T = 0$ (Table 3.7) for this special case. Based on the kinds of subgraphs, the diversity decreases, $\Delta N_S = -14$ and $\Delta T_S = -4$. The significant increase in B_T is an artifact of the way biclique is defined for vertex-colored graphs (section 3.2.9). (If the colors are ignored, then $\Delta B_T = \Delta C_T = 0$.)

The methylene Wittig reagent R_3PCH_2 is a unique one that has the same complexity as the by-product R_3PO, when they are abstracted as hydrogen-suppressed molecular graphs. A more typical Wittig reaction, which gives the same products, is that of the three-carbon aldehyde, acrolein (○=•–•=•), and the three-carbon Wittig reagent (not counting Ph groups), allylidenetriphenylphosphorane (®$_3$Ⓟ=•–•=•). Molecular complexity decreases in this case, $\Delta N_T = -36$ and $\Delta T_T = -24$, as does diversity, $\Delta N_S = -18$ and $\Delta T_S = -8$ (Table 3.7). In contrast to the previous example, where the complexity of the phosphorane is lower than that of the aldehyde, the complexity calculation in this case is dominated by the Wittig reagent, which synthetic chemists usually take for granted. In connection with reactions involving olefins, it should be noted that the *E*- and *Z*-olefin stereoisomers have the same complexity, since their molecular graphs are isomorphic.

The transition metal-mediated cyclotrimerization of alkynes (Reppe-Vollhardt reaction) is shown for acetylene in Figure 3.8. Molecular complexity escalates rapidly with rings and double bonds, which can be thought of as 2-rings. The relatively high complexity of the product (benzene) results in a significant increase for all the 'total' indices: N_T = 252, T_T = 100, B_T = 22 and C_T = 10 (Table 3.8). Among the "kinds of" indices, only N_S = 21 is considered to be significant. The index B_T is easy to determine, even for large molecules, and it is useful for "back of the envelope" calculations. (The complications in this index caused by coloring can be avoided by focusing on the unlabeled molecular graph, vide supra section 3.2.9.)

Another modern synthetic method catalyzed by transition metals is the Pauson-Khand reaction, which forges an alkene, an alkyne and carbon monoxide into a five-membered ring containing a carbon-carbon double bond (Figure 3.9). For the reaction of ethylene, acetylene and carbon monoxide that gives 2-cyclopentenone, we have N_T = 122, T_T = 66, N_S = 42 and T_S = 11 (Table 3.9). When the vertex colors are ignored,

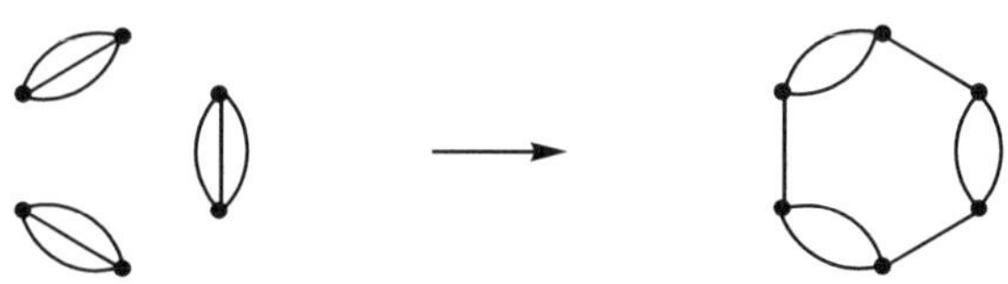

10 **11**

Figure 3.8: Trimerization of acetylene (**10**) to benzene (**11**).

Table 3.8: Trimerization of acetylene (**10**) to benzene (**11**)

Index	**10**	**11**	**Δ**
N_T	9	261	252
N_S	4	25	21
T_T	5	105	100
T_S	2	6	4
B_T	5	27	22
B_S	2	3	1
C_T	5	15	10
C_S	2	2	0

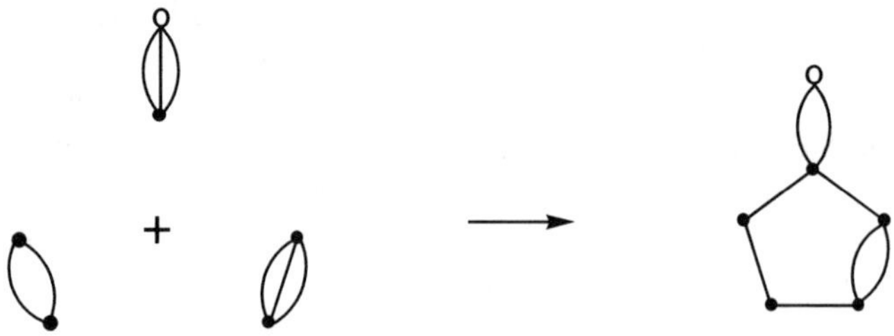

4 **10** **12**

Figure 3.9: Pauson-Khand reaction of ethylene (**4**), acetylene (**10**) and CO, affording 2-cyclopentenone (**12**).

Table 3.9: Pauson-Khand reaction of ethylene (**4**), acetylene (**10**) and CO, affording 2-cyclopentenone (**12**)

Index	4	10	CO	4+10+CO	12	Δ
N_T	5	9	9	23	145	122
N_S	3	4	5	8	50	42
T_T	4	5	5	14	80	66
T_S	2	2	3	4	15	11
B_T	4	5	5	14	8	–6
B_S	2	2	3	3	3	0
C_T	4	5	2	11	12	1
C_S	2	2	2	3	3	0

= 13 is also significant. The total number of atoms in this example is the same as for the Aldol, Diels-Alder and Reppe-Vollhardt reactions discussed above; however, the number of heteroatoms is not the same in every case. As it is independent of coloring, N_T is useful in such cases. Based on N_T, the Pauson-Khand reaction is more similar to alkyne trimerization than to the Aldol or Diels-Alder reactions, in large part because three molecules are joined rather than two (section 3.3.4).

Figure 3.10 shows the conjugate addition of cuprates (House-Whitesides reaction) to 2-cyclopentenone (previous example). For the sake of simplicity, the starting cuprate is treated as the monomer, the aggregation state in tetrahydrofuran solvent [62], rather than the dimer, which is present in ether. The changes in complexity indices for the reaction of 2-

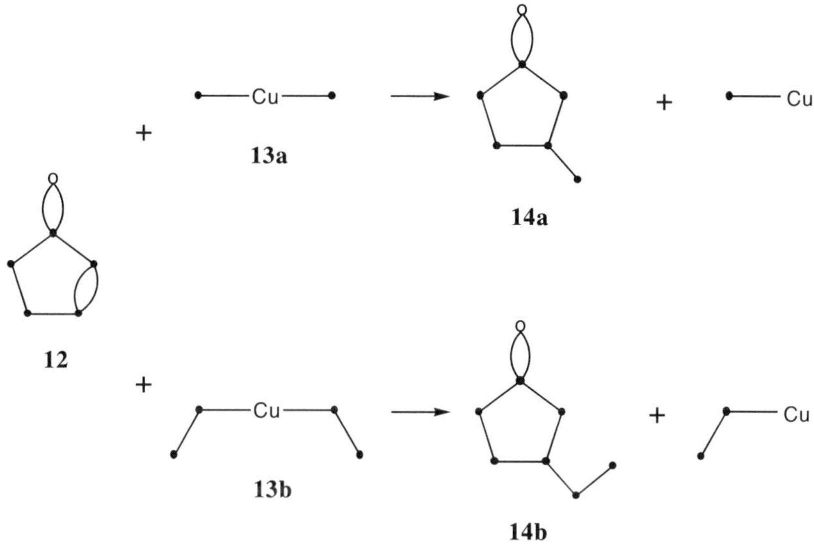

Figure 3.10: Conjugate alkylation of 2-cyclopentenone (**12**) with cuprates (**13a**, methyl; **13b**, ethyl), affording 3-alkylcyclopentanones (**14a**, methyl; **14b**, ethyl).

Table 3.10: Conjugate alkylation of 2-cyclopentenone (**12**) with cuprates (**13a**, methyl; **13b**, ethyl), affording 3-alkylcyclopentanones (**14a**, methyl; **14b**, ethyl)

Index	12	13a (13b)[a]	14a (14b)	MeCu (EtCu)	Δa(Δb)
N_T	145	6 (15)	122 (170)	3 (6)	–26 (16)
N_S	50	4 (8)	47 (65)	3 (5)	–4 (12)
T_T	80	6 (15)	90 (125)	3 (6)	7 (36)
T_S	15	4 (8)	26 (35)	3 (5)	10 (17)
B_T	8	6 (8)	9 (10)	3 (4)	–2 (–2)
B_S	3	4 (4)	3 (3)	3 (3)	–1 (–1)
C_T	12	3 (7)	13 (15)	2 (4)	0 (0)
C_S	3	2 (3)	3 (3)	2 (3)	0 (0)

[a]The cuprate has a formal negative charge and a Li counterion, which are not shown.

cyclopentenone with the methyl cuprate to give 3-methylcyclopentanone are $\Delta N_T = -26$, $\Delta T_T = 7$, $\Delta N_S = -4$ and $\Delta T_S = 10$ (Table 3.10). The overall complexity change for the reaction includes the usual aqueous work-up, which converts the initial enolate to the ketone. This is an interesting example in which the total number of subgraphs diverges from the total number of trees, since N_T decreases upon going from starting materials to products, whereas T_T increases. The same is true for the corresponding "kinds of" indices, N_S and T_S. The carbon-carbon double bond in the starting material contributes more to N_T and N_S than the new methyl group in the product; however, as the group added becomes more complex, these indices also increase. For example, in the reaction of the ethyl cuprate with 2-cyclopentenone to give 3-ethylcyclopentanone, $\Delta N_T = 16$, $\Delta T_T = 36$, $\Delta N_S = 12$ and $\Delta T_S = 17$.

Reactions that are often considered to be trivial, e.g., the reduction of double bonds, can result in significant changes in complexity. In Figure 3.11 we abstract the catalytic hydrogenation of cyclohexene to cyclohexane. There is a substantial decrease in molecular complexity, $\Delta N_T = -32$ and $\Delta T_T = -15$ (Table 3.11). Based on subgraphs, the decrease in diversity is significant ($\Delta N_S = -10$), but not based on trees ($\Delta T_S = 0$). The complexity of dihydrogen is zero, since its hydrogen-suppressed molecular graph is isomorphic to the null graph (G_0). The catalyst is both a starting material

5 **15**

Figure 3.11: Reduction of cyclohexene (**5**) to cyclohexane (**15**).

Table 3.11: Reduction of cyclohexene (**5**) to cyclohexane (**15**)

Index	**5**	**15**	**Δ**
N_T	69	37	–32
N_S	17	7	–10
T_T	51	36	–15
T_S	6	6	0
B_T	21	18	–3
B_S	3	3	0
C_T	13	12	–1
C_S	2	2	0

and a product, and its complexity cancels out upon subtraction (cf. equation 3.5). In general, it is advisable to avoid steps in which complexity decreases, as they are associated with "excess complexity" (next section). Nevertheless, they are sometimes unavoidable and are justified when they enable the use of a powerful synthetic reaction by facilitating the preparation of a key intermediate or by correcting the resulting functionality when it isn't present in the target.

In Figure 3.12 we examine the [3,3]-sigmatropic rearrangement of *Z*-1,3,5-hexatriene to 1,3-cyclohexadiene. (The *Z*-olefin in the starting material is the usual outcome of the Wittig reaction, vide supra.) The total number of subgraphs increases markedly (Table 3.12), $N_T = 38$, as does the total number of trees, $T_T = 25$, owing to the formation of a new ring. The other indices do not change significantly. A rearrangement that does not create a new ring or a more highly substituted carbon atom usually results in a negligible change in complexity, e.g., the degenerate Cope

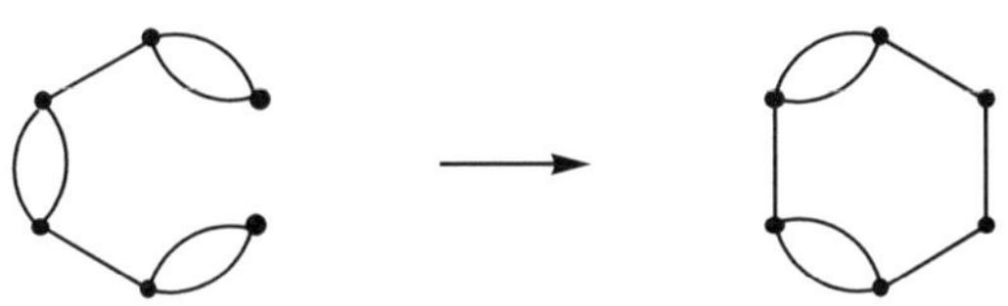

8 **16**

Figure 3.12: [3,3]-Sigmatropic rearrangement of 1,3,5-hexatriene (**8**) to 1,3-cyclohexadiene (**16**).

Table 3.12: [3,3]-Sigmatropic rearrangement of 1,3,5-hexatriene (**8**) to 1,3-cyclohexadiene (**16**)

Index	**8**	**16**	**Δ**
N_T	95	133	38
N_S	19	23	4
T_T	48	73	25
T_S	6	6	0
B_T	22	24	2
B_S	3	3	0
C_T	14	14	0
C_S	2	2	0

Table 3.13: Summary of complexity changes for synthetic reactions

Reaction	ΔN_T (rank)	ΔT_T (rank)	$\Delta\eta$ (rank)	ΔN_S (rank)
Aldol (unlabeled, **1** + **1'**)	26 (5)	22 (6)	4 (3)	12 (3)
Aldol (labeled, **1** + **1***)	16 (6)	14 (7)	1 (5)	11 (5)
Diels-Alder (**3** + **4**)	38 (3)	30 (4)	2 (4)	9 (6)
Wittig (special case, **6a** + **7a**)	0 (8)	0 (9)	0 (7)	–14 (10)
Wittig (typical, **6b** + **7b**)	–36 (11)	–24 (11)	0 (7)	–18 (11)
Reppe-Vollhardt (**10** + **10** + **10**)	252 (1)	100 (1)	12 (1)	21 (2)
Pauson-Khand (**4** + **10** + CO)	122 (2)	66 (2)	6 (2)	42 (1)
House-Whitesides (**12** + Me_2CuLi)	–26 (9)	7 (8)	–2 (9)	–4 (8)
House-Whitesides (**12** + Et_2CuLi)	16 (6)	36 (3)	–2 (9)	12 (3)
Reduction (**5** + H_2)	–32 (10)	–15 (10)	–3 (11)	–10 (9)
[3,3]-sigmatropic rearrangement (**8**)	38 (3)	25 (5)	1 (5)	4 (7)

rearrangement of 1,5-hexadiene [**60**] has $\Delta C = 0$ no matter which index is used, since the starting material and the product are identical.

Table 3.13 summarizes the values of ΔN_T, ΔT_T, $\Delta\eta$ and ΔN_S for the above reactions. The top two reactions, Reppe-Vollhardt and Pauson-Khand, are the same according to all four indices, with the caveat that the order is reversed for ΔN_S. The two reactions tied for #3 according to ΔN_T, Diels-Alder cycloaddition and [3,3]-sigmatropic rearrangement, are also ranked highly by ΔT_T and $\Delta\eta$ (#4 and #5), but somewhat lower by ΔN_S (#6 and #7, respectively). The bottom three reactions according to ΔN_T, methyl cuprate addition, double bond reduction and typical Wittig reaction (#9, #10, #11), are also among the lowest according to ΔT_T (#8, #10, #11), $\Delta\eta$ (#9, #11, #7) and ΔN_S (#8, #9, #11, respectively). The most robust index N_T is tracked remarkably well by the simplest index η, suggesting that past applications of the latter to synthesis planning are fundamentally sound [7, 38, 39, 63], even though it is much less sophisticated.

3.3.2 Complexity versus step plots and excess complexity

Excess complexity C_x is the "extra" area under a "complexity versus step" plot (*C-s* plot), which can be calculated as the sum of the complexities C_i of the n intermediates between the starting materials and the target (equation 3.8) [63]. (N.B., for integration each step has width $\Delta s = 1$.) It is called "excess" because it is the complexity above the absolute minimum required for an *ideal synthesis*, in which the starting materials are converted quantitatively into the target in a single step. As a heuristic, we have proposed the principle of *minimization of excess complexity* [63], which is justified below. Wender [56] has discussed the "maximization of target-relevant complexity," which is a complementary concept, and Trost [64] has highlighted "atom economy," which also tends to minimize excess complexity. In analogy to Whitlock's ΔS_{m} (vide infra), we define ΔC_{m} in equation 3.9, where ΔC_j ($j = 1, \ldots, k$) is the change in complexity for the jth step and k is the total number of steps in the synthesis.

$$C_x = \Sigma_i C_i \tag{3.8}$$

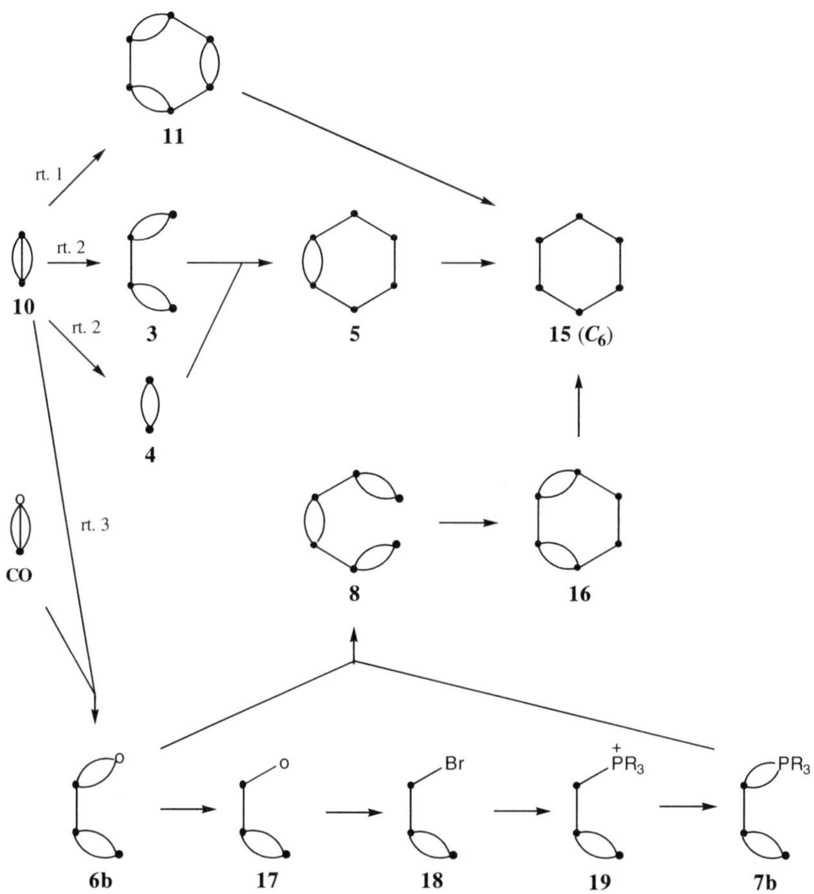

Figure 3.13: Three routes to cyclohexane from acetylene. The counterion for **19** is Br⁻.

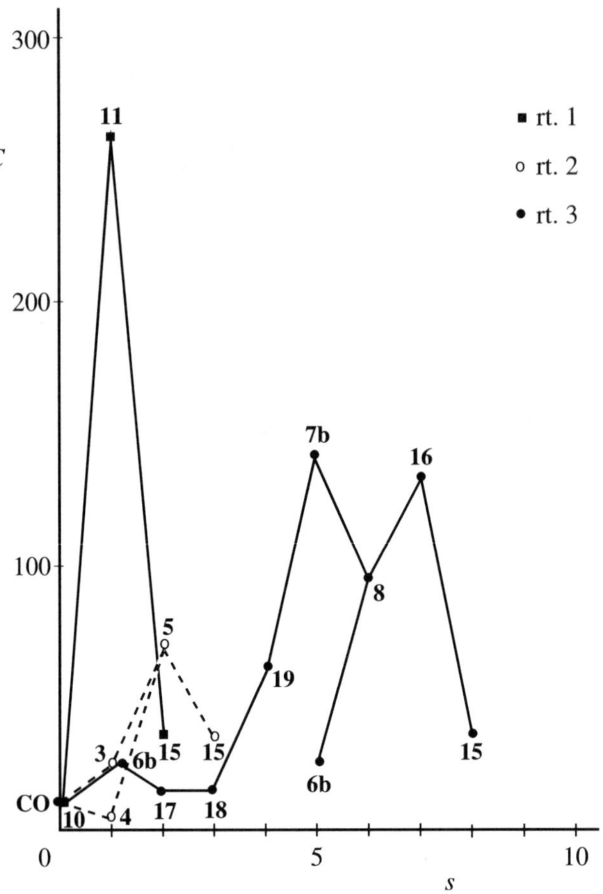

Figure 3.14: Complexity versus step plots for three syntheses of cyclohexane **15**.

$$\Delta C_{\mathrm{m}} = \Sigma_j \left|\Delta C_j\right| / k \tag{3.9}$$

For the three synthetic routes from acetylene to cyclohexane illustrated in Figure 3.13, the complexity versus step plots are presented in Figure 3.14, and Table 3.14 summarizes the data from which they were constructed. For the most valid comparisons of alternative synthetic routes to a given target, the starting materials should be the same, as far as possible. We use $C = N_{\mathrm{T}}$ for the *C-s* plots, since it is a robust index and heteroatoms are not

Table 3.14: Complexity indices N_T and S and derived functions for routes 1–3

Compound[a]	N_T	ΔN_T	S	ΔS	**Functions**
route 1					$C_x = 261$
(acetylene **10**, starting material)	(9)	—	(4)	—	$C_x/n = 261$
benzene **11**	261	252	10	6	$\Delta C_m = 238$
(cyclohexane **15**, product)	(37)	–224	(4)	–6	$S_x = 10$
					$\Delta S_m = 6$
route 2					
(acetylene **10**, starting material)	(9)	—	(4)	—	$C_x = 100$
ethylene **4**	5	–4	2	–2	$C_x/n = 33.3$
butadiene **3**	26	17	4	0	$\Delta C_m = 22.7$
cyclohexene **5**	69	38	6	0	$S_x = 12$
(cyclohexane **15**, product)	(37)	–32	(4)	–2	$\Delta S_m = 1.0$
route 3					$C_x = 489$
(acetylene **10**, starting material)	(9)	—	(4)	—	$C_x/n = 69.9$
(CO, starting material)	(9)	—	(5)	—	$\Delta C_m = 36.7$
(R_3P:, reagent)	(19)	—	(1)	—	$S_x = 34$
(MeLi, reagent)	(3)	—	(1)	—	$\Delta S_m = 1.7$
acrolein **6b**	26	8	5	–4	
allyl alcohol **17**	16	–10	3	–2	
allyl bromide **18**	16	0	3	0	
phosphonium bromide **19**	62[b]	27	4[b]	0	
phosphorane **7b**	141	79[c]	5	2[c]	
1,3,5-hexatriene **8**	95	–36[d]	6	0[d]	
1,3-cyclohexadiene **16**	133	38	8	2	
(cyclohexane **15**, product)	(37)	–96	(4)	–4	

[a]The complexity of dihydrogen is 0 with any index. In route 3 the 2-propanol used in the reduction of **6b** is considered to be a source of hydrogen, and the acetone produced is assumed to be inert. The complexities of HBr and H_2O are equal, and they cancel out (cf. text). Therefore, these three reactants and two products are not included. [b]Includes the bromide counterion ($N_T = 1$, $S = 1$). [c]By-products are methane ($N_T = 1$, $S = 0$) and LiBr ($N_T = 2$, $S = 2$). [d]By-product is R_3PO ($N_T = 36$, $S = 4$).

involved in two of the routes. The reactions used here are taken from the literature.

Route 1 has two reactions and one intermediate. Acetylene is cyclotrimerized via transition metal catalysis to benzene [65, 66], which is reduced by catalytic hydrogenation to cyclohexane [67]. If the acetylene were an intermediate that had to be prepared, this route would be highly convergent and reflexive (section 3.3.5). Catalysts do not appear in the balanced equations, and therefore their complexities are not part of the calculations. The excess complexity is the molecular complexity of the lone intermediate, benzene: C_x(route 1) = N_T(benzene) = 261.

Route 2 has four reactions and three intermediates, and it can be considered to be convergent (section 3.3.5). The first stage of the synthesis has two parallel reactions, which are catalyzed by transition metals, the preparations of butadiene [66, 68] and ethylene [68] from acetylene. These intermediates are joined in the convergent step by the Diels-Alder reaction to afford cyclohexene [69], which is reduced by catalytic hydrogenation to cyclohexane [67]. The excess complexity is the sum of the complexities of the three intermediates: C_x(route 2) = N_T(ethylene) + N_T(butadiene) + N_T(cyclohexene) = 5 + 26 + 69 = 100.

Route 3 is the longest, eight reactions and seven intermediates. It has seven commercial starting materials or reagents, viz. acetylene, CO, HBr, triphenylphosphine (abstracted as R_3P:), MeLi, 2-propanol (hydrogen source) and dihydrogen. Transition metal-catalyzed reductive carbonylation of acetylene with CO gives acrolein [70], and half of it is reduced (2-propanol/catalyst) to allyl alcohol [71], which is converted to allyl bromide with HBr [72]. The bromide is treated with triphenylphosphine to give the phosphonium salt [73], and this precursor is converted with MeLi to the phosphorane, which is allowed to react with the second half of the acrolein. Both reactants in the key, Wittig reaction have the same number of atoms (not including R), and the synthesis can be considered convergent. (Like most convergent routes, it has linear segments.) It also has an element of reflexivity (section 3.3.5), since acrolein is used twice. The *Z*-1,3,5-hexatriene from the Wittig reaction is cyclized via a [3,3]-sigmatropic rearrangement to 1,3-cyclohexadiene [74], which is reduced by catalytic hydrogenation to cyclohexane [67]. When calculating C_x, we add N_T for acrolein once, since the half that is converted into the Wittig reagent can be considered to be an intermediate, and the half that is used as a reactant in the Wittig reaction can be considered to be a starting material. Therefore, the excess complexity is C_x(route 3) = N_T(acrolein) + N_T(allyl alcohol) + N_T(allyl bromide) + N_T(phosphonium) + N_T(phosphorane) + N_T(hexatriene) + N_T(cyclohexadiene) = 26 + 16 + 16 + 62 + 141 + 95 + 133 = 489.

The complexities of allyl alcohol and allyl bromide are the same no matter which index is used, as one unique label is substituted for another in their molecular graphs. The reagent for this conversion is HBr [72], which has the same complexity as the by-product, H_2O, when they are abstracted as hydrogen-suppressed graphs; consequently, $\Delta C = 0$ for this reaction, no matter which index is used. Such functional group interchanges contribute to excess complexity without building up the skeleton, but they are often necessary to adjust reactivity.

The three routes can be evaluated by using old and new heuristics. Route 3 is much longer than the other two and has considerably higher excess complexity. Consequently, it can be eliminated from consideration, even though it has elements of convergence and reflexivity (section 3.3.5). Routes 1 and 2 are both atom economical, and overall they are competitive. Whereas route 1 has higher excess complexity than route 2, it has fewer steps. The decision as to which of these two routes to try first would be based on other considerations, e.g., the unactivated Diels-Alder reaction in route 2 is not facile [69], and the competing dimerization of butadiene is problematical [75]. All things considered, route 1 would be the preferred candidate for initial investigation.

Complexity versus step plots have characteristic profiles. For example, in some cases there is a "peak," a burst of complexity that substantially overshoots the complexity of the target. Such plots tend to be for convergent syntheses (section 3.3.5). Subsequent steps bring the complexity back down to that of the target, or better still the burst occurs in the last step and the peak is the target. Sometimes the peak occurs immediately and other times after many steps, depending on how long it takes to set the stage for a powerful construction reaction. Other syntheses have plots in which the molecular complexity increases incrementally until the target is reached. Such syntheses tend to be linear (section 3.3.5), and they are usually longer than convergent ones. Of course, these examples are archetypes, and everything in between is possible. Typical situations were illustrated by several syntheses of modhephene in the communication that introduced complexity versus step plots [63]. Numerous *C-s* plots were graphically illustrated in the extensive article by Chanon et al. [76], and Serratosa [77] used *C-s* plots to analyze several syntheses of dodecahedrane, projected or accomplished.

All three routes in Figure 3.13 have complexity versus step plots (Figure 3.14) that exhibit a peak significantly higher than the complexity of the target. In route 1 it occurs immediately, as no functional group modification is necessary, and the subsequent reduction gives the product directly. In route 2 there is a brief set-up to prepare the reactants for the key construction reaction (Diels-Alder cycloaddition), and the difference between the peak and target complexities (ΔC for reduction) is not as large. In route 3 the peak occurs late in the synthesis, since the key intermediate (Wittig reagent) must be prepared via a series of steps. The peak in this case is a multiple one, and it goes from "Mount Everest" (Wittig reagent) to "K-2" (1,3-cyclohexadiene) with a pass (1,3,5-hexatriene) in between. The final reduction reaction has ΔC between the other two.

Whitlock [44] has introduced an empirical measure of complexity S for synthetic analysis, as embodied in equation 3.10, where R is the number of rings, U is the number of (non-aromatic) unsaturations, A is the number of asymmetric centers and H is the number of heteroatoms. (N.B., equation 3.10 is our expression for his formulation, which was written in "computerese." In his scheme a triple bond is counted as two double bonds, versus three for our index η.) The choice of coefficients r, u, a and h (cf. next section) is defended as 'chemically reasonable,' which is another application of 'intuition.' This index is used to construct complexity versus step plots, which are analyzed in terms of ΔS_{m} (equation 3.11), the average of $|\Delta S_j|$ for all k steps, where ΔS_j ($j = 1, \ldots, k$) is the change in S for the jth step. In analogy to C_x, we define S_x as the excess complexity based on S of a synthesis with n intermediates, calculated by using equation 3.12, where S_l is the S-value for intermediate I_l ($l = 1, \ldots, n$).

$$S = rR + uU + aA + hH = 4R + 2U + 2A + H \tag{3.10}$$

$$\Delta S_{\mathrm{m}} = \Sigma_j |\Delta S_j| / k \tag{3.11}$$

$$S_x = \Sigma_l S_l \tag{3.12}$$

The values of S, ΔS_{m} and S_x are included in Table 3.14, along with C, ΔC_{m}, C_{x} and C_x/n, the average excess complexity for a synthesis with n intermediates ($C = N_{\mathrm{T}}$ here). While they are not usually counted [44], aromatic unsaturations must be included when calculating S-values for route 1, as the intermediate is benzene. For comparison with the values based on N_{T} (261, 100, and 489), the excess complexities for routes 1 to 3 using $C = T_{\mathrm{T}}$ are 105, 72 and 299, respectively. Using $C = \eta$, they are 15, 16 and 62, and using $C = N_{\mathrm{S}}$, they are 25, 28 and 145, respectively. Based on $C = N_{\mathrm{T}}$ or T_{T}, the order of routes is C_x(route 2) < C_x(route 1) < C_x(route 3). Excess S (10, 12 and 34) gives a different order, S_x(route 1) < S_x(route 2) < S_x(route 3), which is tracked by C_x based on $C = \eta$ or N_{S}. The trends in C_x/n and ΔC_{m} parallel those in ΔS_{m} (Table 3.14), viz. ΔS_{m} (route 2) < ΔS_{m} (route 3) < ΔS_{m} (route 1). All of the functions identify the best synthesis as either route 1 or route 2, which was the conclusion based on heuristics (vide supra).

Which of these quantities correlates best with overall yield? Since the probability of yield-decreasing side reactions tends to increase with the complexities of the reactants and products of all the reactions in a route, the total complexity is a better indicator of probable overall yield than the functions based on complexity changes (ΔC_{m} or ΔS_{m}). (They may be useful for measuring the state of the art, cf. section 3.3.1.) The use of excess complexity C_x instead of total complexity C(total) is explained below. Which complexity index is optimal for C-s plots and C_x is an open question that requires more study. One advantage of N_{T} for yield prediction is the fact that it does not increase with heteroatoms, which generally decrease synthetic complexity (section 3.3.4).

Halfway through a reaction $\mathrm{A} \rightarrow \mathrm{B}$, there is a 1:1 mixture of reactant A and product B, and [A] = [B]. Up to that point, [A] > [B], and after it, [A] < [B]; nevertheless, the integrated concentrations over the whole reaction are equal. If we assign a complexity contribution of ½C(A) + ½C(B) to a reaction that converts A into B, then the total complexity of a route from starting material S to target T with n intermediates I_i is given by C(total) = [½C(S) + ½$C(\mathrm{I}_1)$] + … + [½$C(\mathrm{I}_{i-1})$ + ½$C(\mathrm{I}_i)$] + [½$C(\mathrm{I}_i)$ + ½$C(\mathrm{I}_{i+1})$] + … + [½$C(\mathrm{I}_n)$ + ½C(T)], which simplifies to C(total) = ½C(S) + Σ_i $C(\mathrm{I}_i)$ + ½C(T). Each intermediate I_i (i = 1, …, n) has a contribution of ½$C(\mathrm{I}_i)$ from the reaction that makes it and another ½$C(\mathrm{I}_i)$ from the reaction that converts it into the next intermediate (or target) for a total contribution of $C(\mathrm{I}_i)$. The complexity of an ideal synthesis that converts S into T in one step is C(ideal) = ½C(S) + ½C(T); the excess complexity over this unavoidable minimum is given by C_x = C(total) – C(ideal) = Σ_i $C(\mathrm{I}_i)$ = Σ_i C_i, where $C_i = C(\mathrm{I}_i)$ is the complexity of intermediate I_i (cf. equation 3.8).

Beyond the fact that very few targets can be prepared in a single step, excess complexity is unavoidable, owing to the need for functional groups to activate bond-forming processes. Usually, these functional groups are not present in the target and must be modified or removed. In the course of such operations, there are inevitably side reactions, which decrease the overall yield. Many of the reactions discussed above illustrate the invaluable role played by transition metal catalysts in modern synthetic chemistry [78]. They help minimize excess complexity because auxiliary functional groups are usually not needed. Furthermore, they can enable more than two starting materials to be incorporated in a single step (cf. sections 3.3.1 and 3.3.4). As a general rule, minimizing excess complexity tends to maximize overall yield. This "two dimensional" heuristic is an improvement over the classical minimization of the number of steps.

3.3.3 Synthetic complexity

In our first communication on complexity [4], we did not distinguish between molecular complexity and synthetic complexity, which we have tried to do in subsequent papers [7, 38, 39]. For example, six-membered carbocyclic compounds have been relatively easy to prepare because of well-developed synthetic methods such as the Diels-Alder reaction, Robinson annulation and Birch reduction [60]. For this reason the synthetic complexity of compounds containing six-membered rings is considered to be low. The synthetic complexities of five- and seven-membered carbocycles are higher, since they are not as accessible at the current state of the art. However, this situation is changing, as continuing developments in the Pauson-Khand [79] and Weiss-Cook [80] reactions are lowering the synthetic complexity of the former, and Wender's vinylcyclopropane-olefin cycloaddition [81] may do the same for the latter. The molecular complexities of these ring systems are not changing, since their structures are not changing. What is evolving is the state of the art of synthetic chemistry, which largely determines synthetic complexity.

The repertoire of synthetic reactions is being enhanced by new multicomponent annulations [80, 82, 83] and tandem processes [84], which reduce the synthetic complexity of specific classes of compounds and often result in large overall increases in molecular complexity. Many of the former are "wild cards," as they give spectacular results in special cases, but are not broadly applicable. Some reasonably general ones are the Weiss-Cook reaction [80], the Posner four-component annulation [82] and the Ugi four-component condensation [83]. Transition metals can mediate multistep synthetic reactions, e.g., the Pauson-Khand and Reppe-Vollhardt reactions bring together three starting materials by incorporating one at a time into growing transition metal complexes.

Whitlock [44] recognized the distinction between molecular complexity and synthetic complexity, and he realized that S (equation 3.10) is not global, i.e., it lacks long-range interactions: "First, it is linear in structure and not, in any reasonable sense, recursive or context sensitive. ... Second, S is a function of structure and not the synthetic reactions available." He also perceived that S is subtly influenced by synthetic complexity: "We conclude that S both reflects our biases concerning synthetic complexity and reveals distinct differences between syntheses of even the same target molecule." As far as our indices are concerned, N_T is global, but η is not.

The biases enter when the coefficients are assigned in equation 3.10. Why is r twice u? It follows from the perception that rings are harder to make than double bonds at the current state of the art. Why are asymmetric centers considered a source of complexity? The modest value assigned to a reflects the fact that chiral molecules are still a significant synthetic challenge, although less of one than in the past. In 1956, Woodward [59] noted, "Erythromycin, with all our advantages, looks at present quite hopelessly complex, particularly in view of its plethora of asymmetric centers." By 1976, when the author was a graduate student at Harvard, both the Corey [85] and Woodward [86] groups were closing in on the total synthesis of erythromycin, which has eighteen asymmetric centers. Why is h low? Heteroatom-carbon bonds are generally easier to form than carbon-carbon bonds at the current state of the art [87]. Whitlock's coefficients reflect that state of the art of synthetic chemistry, and they must change with it.

One application of an intrinsic complexity index is to guide the parameterization of schemes involving linear combinations of complexity factors (section 3.2.4). For example, the effect of replacing a carbon-carbon single bond in hexane with a double bond to give hexene is approximately the same as forming a new ring to give cyclohexane. Thus, N_T increases from 21 for hexane to 31 for 1-hexene, 37 for 2-hexene and 39 for 3-hexene, and $N_T = 37$ for cyclohexane. Consequently, for an index of molecular complexity, we would assign the same coefficient to R and U in equation 3.10, e.g., $r = u = 4$, whereas for an index of synthetic complexity, $r > u$, as in equation 3.10. The values of N_T and N_S are the same for both enantiomers of a chiral molecule. Indeed, this is the case for all the topological indices discussed herein, and it can be argued that a mixture of enantiomers is more complex than either pure stereoisomer. Consequently, for an index of molecular complexity, we would set $a = 0$, and for synthetic complexity $a > 0$, as in equation 3.10. The introduction of a heteroatom does not change N_T; consequently, it does not give any guidance as far as h is concerned. Nevertheless, for molecular complexity we would assign h a higher value than for synthetic complexity (cf. previous paragraph), e.g., if $h = 1$ for the former, then $h = 0$ for the latter.

3.3.4 Strategic bond disconnections

An effective method for planning a synthesis is retrosynthetic analysis [87–89], in which one starts with the target molecule and works backwards, breaking bonds until relatively simple starting materials are obtained. (Sometimes it is necessary to make bonds or rearrange them for strategic reasons.) A retrosynthetic *transform* is conceptually the exact reverse of a synthetic reaction and vice versa. Then, the synthesis plan is the microscopic reverse of the retrosynthetic scheme. It is possible to design hypothetical transforms that do not correspond to actual synthetic reactions, and this may aid in the design of new reactions.

A *disconnection* is the minimal subset of transform bonds, which contains those that are broken or formed and the shortest paths (if any) joining their termini. By minimal we mean it does not include ancillary changes in bonding, e.g., in a π system. Ideally, a disconnection simplifies the molecular skeleton by breaking it into two or more pieces or by reducing the number of rings in it. For example, the disconnection corresponding to the Diels-Alder transform breaks two σ bonds separated by one σ bond in a six-membered ring, but the π bond that is broken in this transform is not included, as it does not contribute to skeletal simplification. For any given disconnection there may be several transforms that contain the set of broken bonds, i.e., several synthetic reactions from which to choose. The transforms that accomplish a given disconnection can be very different and difficult to compare as far as complexity is concerned. By using disconnections, we can focus on the underlying topological changes in a retrosynthetic scheme.

The first transform that provides major simplification usually sets the course of a synthesis. Often, it does not correspond to the first transform in a retrosynthetic scheme, as adjustments to the skeleton or functional groups must be made before a powerful transform can be used. (An analogy is the opening series of moves in a game of chess.) The bonds broken in the key simplifying transform are called the *strategic bonds*, and the corresponding disconnection is a *strategic bond disconnection*. Examples of transforms that give large decreases in molecular complexity in the retrosynthetic direction (large increases in the synthetic direction) correspond to the Pauson-Khand and Reppe-Vollhardt reactions (section 3.3.1), and it can be asked, to what extent are their disconnections responsible?

One-, two- and three-bond disconnections of C_6 (cyclohexane) are shown in Figure 3.15. From the values in Table 3.15, the largest simplification ($\Delta N_T = -28$) is for $C_6 \Rightarrow 3P_2$, which breaks three bonds. This disconnection corresponds to the cyclotrimerization of acetylene [65] or ethylene (not developed). The smallest simplification ($\Delta N_T = -16$) is for $C_6 \Rightarrow P_6$, which breaks only one bond. The two-bond disconnections fall

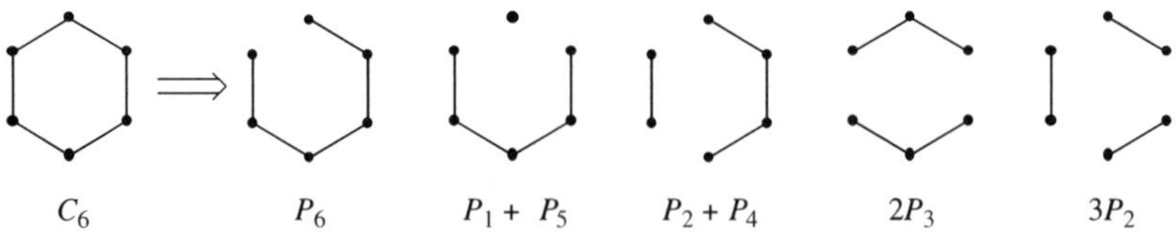

Figure 3.15: Disconnections of cyclohexane (C_6).

Table 3.15: Disconnections of cyclohexane (C_6)

Index	C_6	P_6 (Δ)	P_1+P_5 (Δ)	P_2+P_4 (Δ)	P_3+P_3 (Δ)	3P_2 (Δ)
N_T	37	21 (–16)	16 (–21)	13 (–24)	12 (–25)	9 (–28)
N_S	7	6 (–1)	5 (–2)	4 (–3)	3 (–4)	2 (–5)
T_T	36	21 (–15)	16 (–20)	13 (–23)	12 (–24)	9 (–27)
T_S	6	6 (0)	5 (–1)	4 (–2)	3 (–3)	2 (–4)
B_T	18	15 (–3)	13 (–5)	12 (–6)	12 (–6)	9 (–9)
B_S	3	3 (0)	3 (0)	3 (0)	3 (0)	2 (–1)
C_T	12	11 (–1)	10 (–2)	10 (–2)	10 (–2)	9 (–3)
C_S	2	2 (0)	2 (0)	2 (0)	2 (0)	2 (0)
k_T^{bi}	39	13 (–26)	7 (–32)	5 (–34)	4 (–35)	3 (–36)
k_S^{bi}	7	5 (–2)	4 (–3)	4 (–3)	2 (–5)	1 (–6)
p_T^{bi}	18	8 (–10)	5 (–13)	4 (–14)	4 (–14)	3 (–15)
p_S^{bi}	4	3 (–1)	3 (–1)	3 (–1)	2 (–2)	1 (–3)

in between: $\Delta N_T = -21$ for $C_6 \Rightarrow P_1 + P_5$, $\Delta N_T = -24$ for $C_6 \Rightarrow P_2 + P_4$ (Diels-Alder disconnection), and $\Delta N_T = -25$ for $C_6 \Rightarrow 2P_3$ (bis-allyl disconnection [48]). The same order is obtained by using N_S, T_T or T_S. There are various levels of degeneracy in the other indices based on subgraphs, but the above order is never contradicted.

The same order for the disconnections of C_6 is obtained by using minimal edge biclique covers. Again, the three-bond disconnection gives the largest simplification, $\Delta k_T^{bi}(C_6 \Rightarrow 3P_2) = -36$, and the one-bond disconnection gives the smallest, $\Delta k_T^{bi}(C_6 \Rightarrow P_6) = -26$. The order of two-bond disconnections is also the same: $\Delta k_T^{bi}(C_6 \Rightarrow P_1 + P_5) = -31$, $\Delta k_T^{bi}(C_6 \Rightarrow P_2 + P_4) = -34$ and $\Delta k_T^{bi}(C_6 \Rightarrow 2P_3) = -35$.

The agreement between N_T and k_T^{bi} is satisfying.

Simplification increases with the number of equal pieces that result from a disconnection, e.g., $C_6 \Rightarrow 3P_2$ is better than $C_6 \Rightarrow 2P_3$. For alternate disconnections of a target into a given number of pieces, the maximum simplification occurs when they are equal in complexity, as illustrated by the two-bond disconnections of C_6 (vide supra). These results are also obtained by using indices based on information entropy [26, 27], which explains why $C(\eta)$ works well for strategic bond disconnections [7, 38]. It is not always possible to make the pieces equal, e.g., when there is an odd number of atoms in a ring. Thus, the Pauson-Khand disconnection involves breaking three bonds, but since the five-membered ring cannot be partitioned into three equal pieces, the best we can do is $C_5 \Rightarrow P_2 + P_2 + P_1$. (N.B., the

C_6-O P_6-O1 P_6-O2 P_6-O3

Figure 3.16: Disconnections of oxacyclohexane (C_6-O) to pentanol (P_6-O1), methyl butyl ether (P_6-O2) or ethyl propyl ether (P_6-O3).

Table 3.16: Disconnections of oxacyclohexane (C_6-O) to pentanol (P_6-O1), methyl butyl ether (P_6-O2) or ethyl propyl ether (P_6-O3)

Index	C_6-O	P_6-O1 (Δ)	P_6-O2 (Δ)	P_6-O3 (Δ)
N_T	37	21 (–16)	21 (–16)	21 (–16)
N_S	18	11 (–7)	13 (–5)	12 (–6)
T_T	36	21 (–15)	21 (–15)	21 (–15)
T_S	17	11 (–6)	13 (–4)	12 (–5)
B_T	9	7 (–2)	9 (0)	9 (0)
B_S	4	3 (–1)	4 (0)	4 (0)
C_T	10	10 (0)	9 (–1)	9 (–1)
C_S	3	3 (0)	3 (0)	3 (0)
N_T(lpe)[a]	103	39 (–64)	51 (–52)	57 (–46)
N_S(lpe)[a]	44	23 (–21)	31 (–13)	30 (–14)

[a]Two lone pairs of electrons on oxygen are represented by loops. They are not shown in Figure 3.16.

oxygen of CO is not involved in the disconnection, although it is essential for the reaction.)

Figure 3.16 shows the possible one-bond disconnections of a six-membered ring that contains one heteroatom, e.g., oxacyclohexane (C_6-O). The number of kinds of subgraphs decreases from $N_S = 18$ for C_6-O to $N_S = 11$ for pentanol (P_6-O1), $N_S = 13$ for methyl butyl ether (P_6-O2) and $N_S = 12$ for ethyl propyl ether (P_6-O3). The corresponding changes in diversity are $\Delta N_S = -7, -5$ and -6, respectively (Table 3.16). (N.B., $\Delta N_T = -16$ for all three disconnections.) By including the two lone pairs of electrons (lpe) on oxygen, we have ΔN_T (lpe) = –64, –52 and –46, and ΔN_S (lpe) = –21, –13 and –14, respectively. Based on both of these quantities, as well as ΔN_S [39], breaking the heteroatom-carbon bond results in the greatest simplification. According to the heuristic rules for strategic bonds [87], this is the best one-bond disconnection. The reason is the relative ease of formation of heteroatom-carbon bonds with current synthetic methodology. Our analysis gives the same result, but is purely topological.

Corey and Cheng [87] have posited, "The existence of alternative paths through a molecular skeleton as a consequence of the presence of cyclic subunits gives rise to a *topological complexity* which is proportional to the degree of internal connectivity. Topological strategies are those aimed at the retrosynthetic reduction of connectivity. ... The ordering of strategic disconnections is largely dictated by the mix of strategies used to guide retrosynthetic analysis." Based on the above description, 'topological complexity' is clearly intrinsic. Our work on molecular complexity provides ways to measure the degree of connectivity and thus to order disconnections. Furthermore, double bonds ("2-rings") also contribute to alternate paths, and complexity (N_S, N_T) increases in the series, cyclohexane (7, 37) < cyclohexene (17, 69) < 1,3-cyclohexadiene (23, 133) < benzene (25, 261), where N_T approximately doubles with each new double bond, but N_S levels off markedly after cyclohexene. Therefore, N_T is a good index of topological complexity, but N_S is not useful in this regard.

3.3.5 Linear, convergent and reflexive syntheses

In a *linear synthesis* the intermediates are built up by sequentially adding relatively small molecules, one at a time, until the target is reached. For example, a 27-step linear synthesis of phorboxazole B via "complex Aldol reactions" has been reported [61]. Another example is the synthesis of polypeptides by the Merrifield solid phase method [60], which involves the repeated addition of suitably protected and activated amino acids to the growing chain, anchored to a polymer bead. This technology has been adapted to oligonucleotide synthesis [60].

In a *convergent synthesis* two molecules of approximately the same size are joined in the key step, which forms the strategic bonds (previous section). Synthesis digraphs for linear and convergent routes to a target are shown in Figure 3.17. Since the arcs are understood to point in the forward direction, the arrowheads are often omitted, and a synthesis graph is used to represent the synthetic plan (vide infra). According to the established heuristic [63, 88, 89], a convergent synthesis is more efficient than a linear one of the same target. However, a heuristic is not a law of nature, and there are exceptions. For example, there have been convergent syntheses of peptides that are not as efficient as the linear Merrifield synthesis [7]. It is the reflexive nature of the latter that is responsible for its great efficiency (vide infra).

Convergent syntheses usually have linear segments. For example, Vetter et al. [90] reported "a versatile solid-phase synthesis of *N*-linked

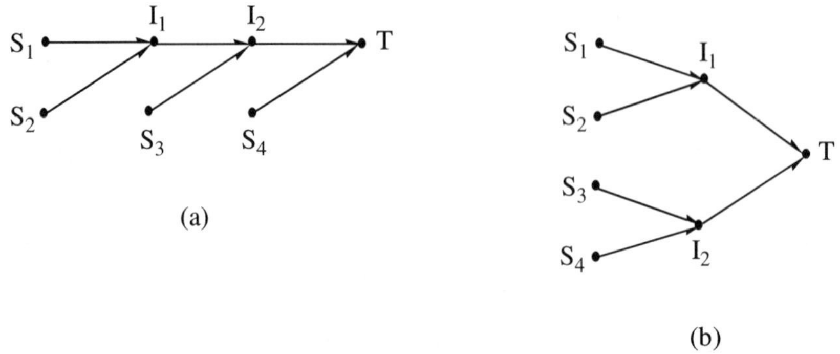

Figure 3.17: Linear (a) and convergent (b) synthesis digraphs for the construction of target T from starting materials S_1, S_2, S_3 and S_4 via intermediates I_1 and I_2.

glycopeptides," and their introduction is instructive: "The discovery of novel glycopeptide ligands for relevant biological receptors has been greatly hindered by the difficult syntheses of these complex compounds. A convergent approach to glycopeptide assembly would improve their accessibility, but few studies have addressed this problem." In the convergent step of the synthesis, a peptide is condensed with an oligosaccharidylamine, both of which are prepared via the linear addition of subassemblies. Boger et al. have illustrated the differences between convergent solution phase and linear solid phase syntheses of combinatorial libraries [91]. Some interesting examples of convergent syntheses are multidrug-resistance reversing agent hapalosin [92], polyamine spider toxin NSTX-3 [93], morpholino-linked ribozymes [94], a 128-Å "molecular wire" [95] and metallodendrimers [96].

Hendrickson [88] has discussed the reasons for the heuristic that convergent syntheses are more efficient than linear ones, i.e., the maximization of convergency: "The qualitative basis for this economy lies in the idea that when a reaction is carried out on an intermediate, it usually involves only one or two of the synthons that make up the intermediate so that the other, uninvolved synthons comprising the intermediate are subjected to needless waste from yield loss in the reaction. Indeed the functionality present on the uninvolved synthons may contribute to yield loss through unwanted side reactions. Also the manipulated weights per step (TW), and hence the time required, are lower in a convergent plan in which fewer uninvolved synthons are carried along in each step (this is true even with 100% yields)." In this context, a synthon is a part of an intermediate or target that is derived from a single starting material.

The synthesis plans shown in Figure 3.17 are trees, since they do not have multiple lines or cycles. Therefore, each line is a *bridge*, meaning its removal results in a disconnected graph, whether the plan is linear or convergent. Such synthesis plans are vulnerable [37]: if any one reaction fails, then the synthesis cannot be completed. As Warren [89] cautioned, "Convergent syntheses are better than linear ones only if all other things are equal. There is no magic about convergence and a bad step can be just as disastrous here as elsewhere." Consequently, the more multiple lines and cycles in the synthesis graph, i.e., the more complex the graph, the better the chances of success.

An important parameter is the maximum number of line-disjoint paths from starting material S to target T, $p_{ld}(S, T)$, which according to Menger's Theorem, is equal to the minimum number of lines that must be removed to make it impossible to transit from S to T [37]. (N.B., line-disjoint paths can share points, but not lines.) In Figure 3.18 the maximum numbers of line-disjoint paths between the starting materials and the target are $p_{ld}(S_1, T) = p_{ld}(S_2, T) = p_{ld}(S_3, T) = 2$ and $p_{ld}(S_4, T) = 3$. Therefore, the failure of two reactions can scuttle this synthesis, when they are both between S_1, S_2 or S_3 and T. At least three reactions must fail between S_4 and T for the plan to fail.

A *reflexive synthesis* is one in which there is symmetry in the synthesis graph. At one extreme, an intermediate merely appears twice, e.g., acrolein in route 3 above (section 3.3.2). At the other extreme, every intermediate is the same (vide infra). A reflexive synthesis results in greater economies of time and effort than the corresponding non-reflexive one, whether it is linear or convergent [38–40]. To demonstrate the efficiency of the reflexive

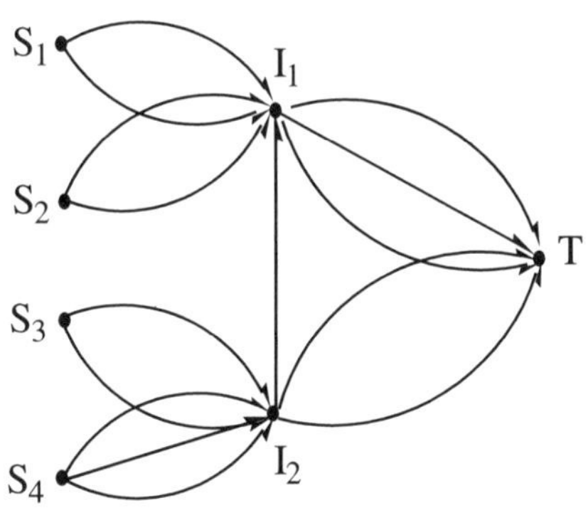

Figure 3.18: Complex synthesis digraph.

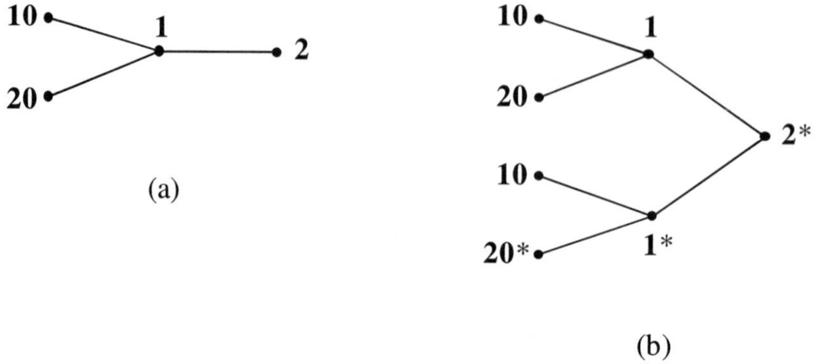

Figure 3.19: Reflexive (a) and non-reflexive (b) synthesis graphs, involving the Aldol reaction, where **1** = acetaldehyde, **1*** = oxygen-labeled acetaldehyde, **2** = aldol, **2*** = oxygen-labeled aldol, **10** = acetylene, **20** = water and **20*** = oxygen-labeled water.

synthesis plan, it can be compared to the non-reflexive plan that is the same in every respect save one, namely the intermediates that are identical in the former are different in the latter. The reactions must be the same in both cases, as far as possible. Our approach is to compare the synthesis of a compound with the same synthesis of an isotopically labeled analog. By judicious choice of labeled atom, any differences in reactivity can be minimized. The example of unlabeled versus labeled usnic acid was used in the communication that introduced the concept of reflexivity [40].

The Aldol reactions of natural abundance and oxygen-labeled acetaldehyde were presented in Figure 3.5. If we prepare these compounds from acetylene and water or oxygen-labeled water [97], then the acetaldehyde isotopomers become intermediates, as shown in the corresponding synthesis graphs (Figure 3.19), which compare the reflexive synthesis of unlabeled aldol to the non-reflexive synthesis of labeled aldol. It is easy to see how the amounts of time and effort required for one person to execute the former are approximately two-thirds of the latter, as two reactions must be run instead of three. This leads to the principle of maximization of reflexivity.

A convergent synthesis is also reflexive when the molecules joined in the key step are identical, e.g., the symmetric Pinacol coupling discussed in the next paragraph. A synthesis that is convergent and reflexive is more economical than an alternative route to the same target that is only convergent. When there is symmetry in the target molecule, a disconnection

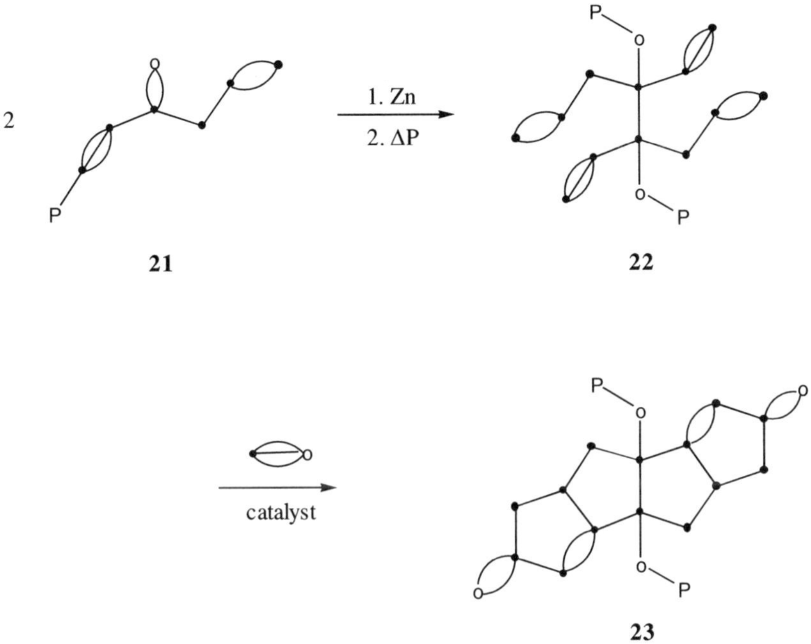

Figure 3.20: Pinacol coupling (**21** → **22**) and Pauson-Khand reaction (**22** → **23**). P is any protecting group, and ΔP indicates a change in protecting groups.

corresponding to a reflexive synthesis is usually obvious. However, there are other cases in which there is no symmetry element in the target, but it is nevertheless possible to disconnect it into two or more identical pieces, e.g., carpanone [38] and usnic acid [40]. This kind of reflexive synthesis can be more difficult to discern. An effective approach for either situation is to look for isomorphic subgraphs in the molecular graphs of the target and intermediates on the way to it (section 3.2.4). Examples of both have been given [38, 98]. Whether or not there is symmetry in the target molecule, there is symmetry in the synthesis graph [40].

The Cook group has reported a reflexive, double-barreled version of the Pauson-Khand reaction (section 3.3.1), which generates six new carbon-carbon bonds and four five-membered rings in a one-pot process [99]. The substrate for carbon monoxide insertion is a molecule that contains two double bonds and two triple bonds. It is prepared from diethyl oxalate via a linear reflexive route that incorporates two equivalents of the same alkene in the form of allylzinc bromide and two equivalents of the same alkyne in the form of LiC ≡ C-TMS. Another synthetic route to a bis(alkene)-bis(alkyne) involves a symmetric Pinacol coupling, which is convergent and reflexive (Figure 3.20). The reflexivity of the Pinacol coupling step is very important for the overall efficiency of the Cook synthesis.

Dramatic examples of reflexive linear syntheses are polymerization reactions which produce high molecular weight polymers, e.g., polystyrene [60]. Styrene monomers are added one at a time to the growing chain until the reaction is terminated. If it is assumed that $(\text{styrene})_n \approx (\text{styrene})_{n-1}$, then each propagation step has the same starting materials (styrene monomer and polymer chain), and the synthesis is highly reflexive. This approximation gets better as the chain grows longer. The symmetry in the polystyrene molecule is translational symmetry [29] rather than the familiar point group symmetry of small molecules. (In all cases one must consider the symmetry of the most symmetric presentation of the molecule.) The Merrifield polypeptide synthesis is reflexive because to a first approximation the amino acids are all treated in the same way, and the *n*-mer of the growing chain is essentially the same as the $(n - 1)$-mer. This reflexive linear synthesis is more efficient than the non-reflexive, convergent synthesis of polypeptides.

The archetypes "linear" and "convergent" are extremes, and everything in between is possible. Moreover, these classifications are somewhat arbitrary, since they depend on the relative sizes of the molecules joined. For example, the Diels-Alder reaction of butadiene and ethylene was called convergent (section 3.3.2), as the difference in size (number of atoms) was not great, and it was necessary to prepare both of them. On the other hand, in terms of complexity N_T, the difference is much larger, and route 2 could be considered linear, especially if the ethylene were purchased. Cook's Pinacol coupling (Figure 3.20) is unequivocally convergent, as the two molecules joined are identical. The Pauson-Khand reaction in Figure 3.9 (section 3.3.1) can be considered convergent, as the three molecules joined are equal in size and similar in complexity. In contrast, the Pauson-Khand reaction in Figure 3.20 involves one starting material (CO) that is much smaller and less complex than the other (14-atom substrate, not counting protecting groups), and it is clearly linear. The reflexivity of this step is trivial, using commercial CO. On the other hand, if the CO were an isotopomer (e.g., ^{11}CO) that had to be prepared via a multistep route, then this step would be convergent, and reflexivity would also be important.

We do not split hairs in these classifications, and this is one place where we defer to intuition, since convergence and reflexivity are heuristics.

3.4 Conclusion

R.B. Woodward admonished [59], "Synthesis must always be carried out by plan, and the synthetic frontier can be defined only in terms of the degree to which realistic planning is possible, utilizing all of the intellectual and physical tools available." We have introduced a number of new intellectual tools: hierarchies of homologous series, indices of complexity and diversity, complexity versus step plots, excess complexity, and reflexivity. The kinds of complexity have been classified as intrinsic and extrinsic. Intrinsic complexity is associated with the number of parts of a system and their interactions or other relationships. Extrinsic complexity is determined by the way a system is constructed, either physically or conceptually. The intrinsic complexity of a molecule is its molecular complexity, and the extrinsic complexity of construction is its synthetic complexity. Symmetry is a simplifying factor for extrinsic complexity, but not for intrinsic complexity.

By mathematically formalizing the concept of homologous series, introducing their hierarchies and clarifying the role of symmetry, we are able to establish that certain graph invariants $I(G)$ are indices of intrinsic complexity, e.g., the number of connections η, the total number of subgraphs N_T, and the total number of minimal edge biclique covers k_T^{bi}. These indices are useful for synthetic analysis, e.g., measuring the changes in molecular complexity for synthetic reactions and retrosynthetic bond disconnections. They also enable us to construct complexity versus step plots for alternate synthetic routes to a target and to calculate their excess complexities. Consideration of symmetry in synthesis graphs leads to the concept of reflexivity, which can afford dramatic economies of time and effort. Our new heuristics, minimization of excess complexity and maximization of reflexivity, have been added to the classic ones, such as the minimization of the number of steps and the maximization of convergence.

Finally, different areas of study have their own criteria of complexity. While we have focused on the meaning of complexity in chemistry [100], our approach is based on a very general mathematical approach, which should be applicable to all the topics mentioned in the introduction, from psychology to physics [16–25]. We suggest that graph theory may provide the common language of complexity.

Acknowledgments

The author thanks C. Rücker for constructive criticism, T. Sommer and C. Zamfirescu for helpful discussions and W.F. Wright for valuable literature searches.

3.5 References

1. *Webster's Ninth New Collegiate Dictionary*, Merriam-Webster, Springfield, MA, 1983.
2. *McGraw-Hill Dictionary of Scientific and Technical Terms*, 3rd edn, S.P. Parker (Ed.), McGraw-Hill, New York, 1984.
3. D. Bonchev, *J. Chem. Inf. Comput. Sci.*, **40**, 934–941 (2000).
4. S.H. Bertz, *J. Am. Chem. Soc.*, **103**, 3599–3601 (1981).
5. S.H. Bertz, *J. Chem. Soc., Chem. Commun.*, 818–820 (1981).
6. S.H. Bertz, *Bull. Math. Biol.*, **45**, 849–855 (1983).
7. S.H. Bertz, *Chemical Applications of Topology and Graph Theory*, R.B. King (Ed.), Elsevier, Amsterdam, 1983, pp. 206–221.
8. S.H. Bertz, *Discrete Appl. Math.*, **19**, 65–83 (1988).
9. D. Minoli, *Atti Accad. Naz. Lincei Rend. Cl. Sci. Fis. Mat. Natur. (Ser. 8)*, **59**, 651–661 (1975).
10. D. Bonchev and O.E. Polansky, *Graph Theory and Topology in Chemistry*, R.B. King and D.H. Rouvray (Eds), Elsevier, Amsterdam, 1987, pp. 126–158.
11. a) D. Bonchev and N. Trinajstić, *J. Chem. Phys.*, **67**, 4517–4533 (1977); see also b) D. Bonchev, *J. Mol. Struct. (Theochem)*, **336**, 137–156 (1995).
12. D. McShea, *Evolution*, **50**, 477–492 (1996).
13. a) M.L. McKinney, *Science*, **243**, 103 (1989); this is a book review of b) J.T. Bonner, *The Evolution of Complexity by Means of Natural Selection*, Princeton University Press, Princeton, NJ, 1988.
14. S.J. Gould, *Natural History*, February 2000, 28–44.
15. C.K. Yoon, *The New York Times*, March 30, 1993, C1.
16. M.M. Waldrop, *Complexity: The Emerging Science at the Edge of Order and Chaos*, Touchstone-Simon & Schuster, New York, 1992.
17. M. Gell-Mann, *The Quark and the Jaguar: Adventures in the Simple and the Complex*, W.H. Freeman, New York, 1994.
18. Y. Bar-Yam, *Dynamics of Complex Systems*, Addison-Wesley, Reading, MA, 1997.

19. G. Nicolis and I. Prigogine, *Exploring Complexity*, Freeman, New York, 1989.
20. R. Badii and A. Politi, *Complexity: Hierarchical Structures and Scaling in Physics*, Cambridge University Press, Cambridge, 1997.
21. J.L. Casti, *Complexification*, Harper-Collins, New York, 1994.
22. *Complexity: Metaphors, Models and Reality*, G.A. Cowan, D. Pines and D. Meltzer (Eds), Addison-Wesley, Reading, MA, 1994.
23. a) *From Simplicity to Complexity in Chemistry—and Beyond* (Part I), A. Müller, A. Dress and F. Vögtle (Eds), Vieweg, Braunschweig, 1996; b) *From Simplicity to Complexity: Information–Interaction–Emergence* (Part II), K. Mainzer, A. Müller and W.G. Saltzer (Eds), Vieweg, Braunschweig, 1998.
24. *Interdisciplinary Approaches to Nonlinear Complex Systems*, H. Haken and A. Mikhailov (Eds), Springer-Verlag, Berlin, 1993.
25. K. Mainzer, *Thinking in Complexity: The Complex Dynamics of Matter, Mind, and Mankind*, 3rd edn, Springer-Verlag, Berlin, 1997.
26. C.A. Shannon and W. Weaver, *The Mathematical Theory of Communication*, University of Illinois Press, Urbana, 1949.
27. A. Mowshowitz, *Bull. Math. Biophys.*, **30**, 175–204 (1968).
28. R. Hinegardner and J. Engelberg, *J. Theor. Biol.*, **104**, 7–20 (1983).
29. J. Rosen, *Symmetry Discovered: Concepts and Applications in Nature and Science*, Cambridge University Press, Cambridge, 1975.
30. a) A.O. Barut, *Symmetries in Science*, B. Gruber and R.S. Millman, (Eds) Plenum, New York, 1980, pp. 39–54; b) *Symmetries in Science II*, B. Gruber and R. Lenczewski (Eds), Plenum, New York, 1986.
31. a) T.-L. Ho, *Symmetry: A Basis for Synthesis Design*, Wiley, New York, 1995; see also b) D.J. Milner, *J. Chem. Technol. Biotechnol.*, **63**, 301–312, (1995).
32. a) W. Thomson (Lord Kelvin), *Popular Lectures and Addresses*, vol. 1 (1889). From the lecture "Electrical Units of Measurement," delivered May 3 (1883); see also b) *The Oxford Dictionary of Quotations*, 5th edn, E. Knowles (Ed.), Oxford University Press, Oxford, 1999, p. 432.
33. F. Harary, *Graph Theory*, Addison-Wesley, Reading, MA, 1969.
34. T.A. McKee and F.R. McMorris, *Intersection Graph Theory*, SIAM, Philadelphia, 1999.
35. K.A. Ross and C.R.B. Wright, *Discrete Mathematics*, 4th edn, Prentice-Hall, Upper Saddle River, NJ, 1999.
36. D.I.A. Cohen, *Basic Techniques of Combinatorial Theory*, Wiley, New York, 1978.

37. S.H. Bertz, *J. Chem. Soc., Chem. Commun.*, 1627–1628 (1986).
38. S.H. Bertz and T.J. Sommer, *Organic Synthesis: Theory and Applications*, vol. 2, T. Hudlicky (Ed.), JAI Press, Greenwich, CT, 1993, pp. 67–92.
39. S.H. Bertz and W.F. Wright, *Graph Theory Notes of New York*, NY Acad. Sci., XXXV, 32–48 (1998).
40. S.H. Bertz, *J. Chem. Soc., Chem. Commun.*, 218–219 (1984).
41. S.H. Bertz and C.M. Zamfirescu, *Commun. Math. Comput. Chem. (MATCH)*, **42**, 39–70 (2000).
42. C.J. Suckling, K.E. Suckling and C.W. Suckling, *Chemistry through Models*, Cambridge University Press, Cambridge, 1978.
43. a) G. Rücker and C. Rücker, *J. Chem. Inf. Comput. Sci.*, **40**, 99–106 (2000); see also b) D. Bonchev, O. Mekenyan and N. Trinajstić, *Int. J. Quantum Chem.*, **17**, 845–893 (1980).
44. a) H.W. Whitlock, *J. Org. Chem.*, **63**, 7982–7989 (1998); see also b) R. Barone and M. Chanon, *J. Chem. Inf. Comput. Sci.*, **41**, 269–272 (2001).
45. a) M. Gordon and J.W. Kennedy, *J. Chem. Soc., Faraday Trans. II*, **69**, 484–504 (1973); see also b) J.W. Essam, J.W. Kennedy, M. Gordon and P. Whittle, *J. Chem. Soc., Faraday Trans. II*, **73**, 1289–1307 (1977).
46. J.-M. Lehn, *Supramolecular Chemistry: Concepts and Perspectives*, VCH, Weinheim, 1995.
47. D.J. Price, J.D. Roberts and W.L. Jorgensen, *J. Am. Chem. Soc.*, **120**, 9672–9679 (1998).
48. S.H. Bertz and T.J. Sommer, *Chem. Commun.*, 2409–2410 (1997).
49. a) H. Wiener, *J. Am. Chem. Soc.*, **69**, 17–20 (1947); see also b) J.R. Platt, *J. Phys. Chem.*, **56**, 328–336 (1952).
50. S.H. Bertz and W.C. Herndon, *Artificial Intelligence Applications in Chemistry* (ACS Symposium Series, vol. 306), T.H. Pierce and B.A. Hohne (Eds), American Chemical Society, Washington, DC, 1986, pp. 169–175.
51. a) D. Bonchev, *SAR & QSAR Environ. Res.*, **7**, 23–43 (1997); see also b) D. Bonchev, *Topological Indices and Related Descriptors*, J. Devillers and A.T. Balaban (Eds), Gordon and Breach, Reading, UK, 1999, pp. 361–401.
52. R.G.A. Bone and H.O. Villar, *J. Comput. Chem.*, **18**, 86–107 (1997).
53. a) G. Rücker and C. Rücker, *J. Chem. Inf. Comput. Sci.*, **41**, 314–320 (2001); for an improved program, see b) G. Rücker and C. Rücker, *J. Chem. Inf. Comput. Sci.*, **41**, 1457–1462 (2001).

54. R.L. Brooks, C.A.B. Smith, A.H. Stone and W.T. Tutte, *Duke Math. J.*, **7**, 312–340 (1940).
55. a) I. Gutman, R.B. Mallion and J.W. Essam, *Mol. Phys.*, **50**, 859–877 (1983); for a comparison of spanning trees with some of the other indices, see b) S. Nikolić, N. Trinajstić, I.M. Tolić, G. Rücker and C. Rücker, chap. 2 herein.
56. a) P.A. Wender, S.T. Handy and D.L. Wright, *Chem. Ind.*, 765–769 (1997); b) P.A. Wender and B.L. Miller, *Organic Synthesis: Theory and Applications*, vol. 2, T. Hudlicky (Ed.), JAI Press, Greenwich, CT, 1993, pp. 27–66.
57. T. Hudlicky, *Chem. Rev.*, **96**, 3–30 (1996).
58. D. Bradley, *New Scientist*, November 15, 40–43 (1997).
59. R.B. Woodward, *Perspectives in Organic Chemistry*, A. Todd (Ed.), Interscience, New York, 1956, pp. 155–184.
60. a) M.B. Smith and J. March, *March's Advanced Organic Chemistry—Reactions, Mechanisms, and Structure*, 5th edn, Wiley, New York, 2000; b) G. Solomons and C. Fryhle, *Organic Chemistry*, 7th edn, Wiley, New York, 2000.
61. D.A. Evans, D.M. Fitch, T.E. Smith and V.J. Cee, *J. Am. Chem. Soc.*, **122**, 10033–10046 (2000).
62. S.H. Bertz, *J. Am. Chem. Soc.*, **112**, 4031–4032 (1990).
63. S.H. Bertz, *J. Am. Chem. Soc.*, **104**, 5801–5803 (1982).
64. a) B.M. Trost, *Science*, **254**, 1471–1477 (1991); b) B.M. Trost, *Angew. Chem. Int. Ed. Engl.*, **34**, 259–281 (1995).
65. a) W. Reppe, N. v. Kutepow and A. Magin, *Angew. Chem. Int. Ed. Engl.*, **8**, 727–733 (1969); b) K.P.C. Vollhardt, *Angew. Chem. Int. Ed. Engl.*, **23**, 539–556 (1984); for recent papers see ref. 66 and c) I.M. Abdelrehim, K. Pelhos, T.E. Madey, J. Eng and J.G. Chen, *J. Phys. Chem. B.*, **102**, 9697–9707 (1998).
66. T.E. Caldwell, I.M. Abdelrehim and D.P. Land, *Langmuir*, **14**, 1407–1410 (1998).
67. M. Castiglioni, S. Deabate, E. Garrone, R. Giordano, B. Onida, G. Predieri and E. Sappa, *J. Cluster Sci.*, **8**, 381–405 (1997).
68. W. Büchele, H. Roos, H. Wanjek and H.J. Müller, *Catal. Today*, **30**, 33–39 (1996).
69. G.M. Whitman, Brit. Patent 616,671 (1949).
70. S.N. Orlik, Y.I. Pyatnitsky and G.G. Girushtin, *Ukr. Khim. Zh.*, **58**, 737–742 (1992).
71. Y. Nagase and T. Katou, *Chem. Lett.*, 436–437 (2000).
72. J.-M. Mas and P. Metivier, Eur. Patent 398,774 (1990).

73. G. Büchi and H. Wüest, *Helv. Chim. Acta*, **54**, 1767–1776 (1971).
74. S. Sakai and S. Takane, *J. Phys. Chem. A*, **103**, 2878–2882 (1999).
75. F.-G. Klärner, B. Krawczyk, V. Ruster and U.K. Deiters, *J. Am. Chem. Soc.*, **116**, 7646–7657 (1994).
76. M. Chanon, R. Barone, C. Baralotto, M. Juillard and J.B. Hendrickson, *Synthesis*, 1559–1583 (1998).
77. F. Serratosa, *Organic Chemistry in Action: The Design of Organic Synthesis*, Elsevier, Amsterdam, 1990, pp. 19–22, 67, 331.
78. a) K. Weissermel and H.-J. Arpe, *Industrial Organic Chemistry*, Wiley-VCH, New York, 1997; b) L.S. Hegedus, *Transition Metals in the Synthesis of Complex Organic Molecules*, 2nd edn, University Science Books, Mill Valley, CA, 1999.
79. N.E. Schore, *Org. React.*, **40**, 1–90 (1991).
80. a) A.K. Gupta, X. Fu, J.P. Snyder and J.M. Cook, *Tetrahedron*, **47**, 3665–3710 (1991); b) S.H. Bertz, J.M. Cook, A. Gawish and U. Weiss, *Org. Synth.*, **64**, 27–38 (1986).
81. P.A. Wender, A.J. Dyckman, C.O. Husfeld, D. Kadereit, J.A. Love and H. Rieck, *J. Am. Chem. Soc.*, **121**, 10442–10443 (1999).
82. a) G.H. Posner, *Chem. Rev.*, **86**, 831–844 (1986); b) G.H. Posner, Q. Wang, B.A. Halford, J.S. Elias and J.P. Maxwell, *Tetrahedron Lett.*, **41**, 9655–9659 (2000).
83. I. Ugi, A. Dömling and W. Hörl, *Endeavor*, **18**, 115–122 (1994).
84. a) T.-L. Ho, *Tandem Organic Reactions*, Wiley, New York, 1992; b) B.M. Trost, *J. Am. Chem. Soc.*, **114**, 791–792 (1992); c) G.M. Williams, S.D. Roughley, J.E. Davies, A.B. Holmes and J.P. Adams, *J. Am. Chem. Soc.*, **121**, 4900–4901 (1999); d) L.F. Tietze, *Chem. Rev.*, **96**, 115–136 (1996); e) J.D. Winkler, *Chem. Rev.*, **96**, 167–176 (1996); f) P.J. Parsons, C.S. Penkett and A.J. Shell, *Chem. Rev.*, **96**, 195–206 (1996); g) A. Padwa and M.D. Weingarten, *Chem. Rev.*, **96**, 223–269 (1996); h) M. Malacria, *Chem. Rev.*, **96**, 289–306 (1996).
85. E.J. Corey, S. Kim, S. Yoo, K.C. Nicolaou, L.S. Melvin, D.J. Brunelle, J.R. Falck, E.J. Trybulski, R. Lett and P.W. Sheldrake, *J. Am. Chem. Soc.*, **100**, 4620–4622 (1978).
86. R.B. Woodward, E. Logusch, K.P. Nambiar, K. Sakan, D.E. Ward, B.-W. Au-Yeung, P. Balaram, L.J. Browne, P.J. Card, C.H. Chen, R.B. Chênevert, A. Fliri, K. Frobel, H.-J. Gais, D.G. Garratt, K. Hayakawa, W. Heggie, D.P. Hesson, D. Hoppe, I. Hoppe, J.A. Hyatt, D. Ikeda, P.A. Jacobi, K.S. Kim, Y. Kobuke, K. Kojima, K. Krowicki, V.J. Lee, T. Leutert, S. Malchenko, J. Martens, R.S. Matthews, B.S. Ong, J.B. Press, T.V. Rajan Babu, G. Rousseau, H.M. Sauter, M.

Suzuki, K. Tatsuta, L.M. Tolbert, E.A. Truesdale, I. Uchida, Y. Ueda, T. Uyehara, A.T. Vasella, W.C. Vladuchick, P.A. Wade, R.M. Williams and H.N.-C. Wong, *J. Am. Chem. Soc.*, **103**, 3215–3217 (1981).

87. a) E.J. Corey and X.-M. Cheng, *The Logic of Chemical Synthesis*, Wiley, New York, 1989, p. 37 (quote); b) E.J. Corey, W.J. Howe, H.W. Orf, D.A. Pensak and G. Petersson, *J. Am. Chem. Soc.*, **97**, 6116–6124 (1975).
88. a) J.B. Hendrickson, *J. Am. Chem. Soc.*, **99**, 5439–5450 (1977); see also b) J.B. Hendrickson, *Accts. Chem. Res.*, **19**, 274–281 (1986).
89. S. Warren, *Organic Synthesis: the Disconnection Approach*, Wiley, Chichester, 1982, p. 352.
90. D. Vetter, D. Tumelty, S.K. Singh and M.A. Gallop, *Angew. Chem. Int. Ed. Engl.*, **34**, 60–63 (1995).
91. D.L. Boger, W. Chai and Q. Jin, *J. Am. Chem. Soc.*, **120**, 7220–7225 (1998).
92. A.K. Ghosh, W. Liu, Y. Xu and Z. Chen, *Angew. Chem. Int. Ed. Engl.*, **35**, 74–76 (1996).
93. I.S. Blagbrough, E. Moya and S.P. Walford, *Tetrahedron Lett.*, **37**, 551–554 (1996).
94. L. Bellon, C. Workman, J. Scherrer, N. Usman and F. Wincott, *J. Am. Chem. Soc.*, **118**, 3771–3772 (1996).
95. J.S. Schumm, D.L. Pearson and J.M. Tour, *Angew. Chem. Int. Ed. Engl.*, **33**, 1360–1363 (1994).
96. W.T.S. Huck, L.J. Prins, R.H. Fokkens, N.M.M. Nibbering, F.C.J.M. van Veggel and D.N. Reinhoudt, *J. Am. Chem. Soc.*, **120**, 6240–6246 (1998).
97. N.D. Lazo, J.L. White, E.J. Munson, M. Lambregts and J.F. Haw, *J. Am. Chem. Soc.*, **112**, 4050–4052 (1990).
98. a) A.B. Smith, III, S.A. Kozmin and D.V. Paone, *J. Am. Chem. Soc.*, **121**, 7423–7424 (1999); b) J.M. Schkeryantz, J.C.G. Woo and S.J. Danishefsky, *J. Am. Chem. Soc.*, **117**, 7025–7026 (1995); c) T.R. Kelly and R.L. Xie, *J. Org. Chem.*, **63**, 8045–8048 (1998); d) C. Li, E. Lobkovsky and J.A. Porco, Jr., *J. Am. Chem. Soc.*, **122**, 10484–10485 (2000); e) A. Ganesan, *Angew. Chem. Int. Ed. Engl.*, **35**, 611–615 (1996); f) R.S. Coleman and E.B. Grant, *J. Am. Chem. Soc.*, **117**, 10889–10904 (1995); g) R.W. Scott, J. Epperson and C.H. Heathcock, *J. Org. Chem.*, **63**, 5001–5012 (1998); h) R.M. Borzilleri, S.M. Weinreb and M. Parvez, *J. Am. Chem. Soc.*, **117**, 10905–10913 (1995); i) A.G.M. Barrett, M.L. Boys and T.L. Boehm, *J. Org. Chem.*, **61**,

685–699 (1996); j) T.S. McDermott, A.A. Mortlock and C.H. Heathcock, *J. Org. Chem.*, **61**, 700–709 (1996).

99. a) S.G. Van Ornum, M.M. Bruendl, H. Cao, M. Reddy, D.S. Grubisha, D.W. Bennett and J.M. Cook, *J. Org. Chem.*, **65**, 1957–1971 (2000); b) H. Cao, S.G. Van Ornum and J.M. Cook, *Tetrahedron Lett.*, **41**, 5313–5316 (2000); c) S.G. Van Ornum and J.M. Cook, *Tetrahedron Lett.*, **37**, 7185–7188 (1996); d) S.G. Van Ornum and J.M. Cook, *Tetrahedron Lett.*, **38**, 3657–3658 (1997).

100. This is contribution 17 in a series on applications of graph theory to chemistry. For paper 16 see S.H. Bertz, *Chem. Commun.*, 2516–2517 (2001). Paper 15 is ref. 41 and paper 14 is ref. 39.

Chapter 4

SHANNON'S INFORMATION AND COMPLEXITY

D. Bonchev

Department of Marine Sciences, Program for Theory of Complex Natural Systems, Texas A&M University, Ford Crockett Campus, 5007 Avenue U, Galveston, Texas 77551, USA

4.1 Introduction

The intuitive idea of complexity as an antonym of simplicity is an inherent one for the people of the twenty-first century. We live in a complex environment and deal with a complex technology. Every day we face the challenging complexity of life at our work and at home. We enjoy the complexities of our education and entertainment. Could, however, the fuzzy idea of complexity be transformed into a rigorously defined scientific notion? Could one quantitatively assess complexity and, if the answer is "yes," why is it needed?

In this chapter, we will try to shed some light on the above-mentioned questions insofar as they refer to the realm of chemistry. However, the approach we adopt is a general one and it could be applied to any system having a "structure," i.e., to any system composed of certain parts or elements united in a single entity by certain relationships. Processes are not excluded from our consideration provided they include mutually related steps, catalytic chemical reactions being a typical example [1–3].

Consider a couple of chemically relevant cases. Two important classes of industrially produced plastics are high-density polyethylene (HDPE) and low-density polyethylene (LDPE). HDPE is composed mainly of linear macromolecules, whereas in LDPE the macromolecules are highly branched (Figure 4.1a). The two types of polyethylene have very different properties: HDPE has good mechanical characteristics and can be used as construction material, whereas LDPE has low toughness but excellent processability and finds applications for packings. It is intuitively clear that the branched LDPE structures are more complex than the linear HDPE ones. Thus, complexity strongly influences the properties of this material.

Another example is provided by molecules containing the same number of benzene rings connected in a ribbon, a two-dimensional array, and a macro ring (Figure 4.1b). This sequence of benzenoid structures can be extended even further to include three-dimensional shapes, such as spheres (fullerenes), cylinders (nanotubes), etc. These classes of benzenoid structures differ in their chemical reactivity and physical properties. Complexity comparisons between these classes can no longer rely on intuition only; a quantitative measure of complexity is required. Such a relevant measure could be very useful in structure-property relationships, enabling the more effective search for new materials.

4.2 The Notion of Entropy in Physics

Clausius introduced entropy in thermodynamics in the mid-nineteenth century to characterize the changes occurring during irreversible processes in systems that do not interact with their surroundings (isolated systems). According to the Second Law of Thermodynamics, such processes (common examples are heat exchange and diffusion) are always associated with an increase in entropy. Half a century later, Boltzmann revealed the statistical character of the Second Law. The only processes allowed to occur spontaneously are those that increase the disorder in systems. Thus, two gases mix irreversibly; they never separate without outside intervention. Similarly, two gases with different temperatures spontaneously equalize their temperature but the process cannot be reversed without some compensation.

A key role in Boltzmann's statistical thermodynamics is played by his famous formula relating the entropy S to the thermodynamic probability W;

(a)

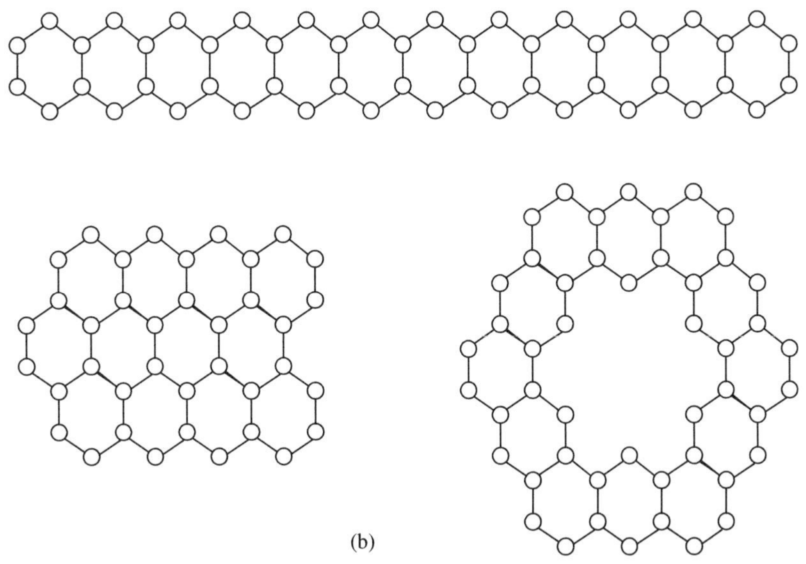

(b)

Figure 4.1: (a)Linear and branched macromolecules characterize high-density polyethylene (HDPE) and respectively low-density polyethylene (LPDE); (b) A ribbon, a two-dimensional array, and a macro ring of benzene rings in benzenoid hydrocarbons.

$$S = k \ln W \tag{4.1}$$

where k is a proportionality constant. Thermodynamic probability is equal to the number of microstates in a system macrostate, characterized by a constant total energy. For a system having N particles distributed such that N_1 of them have energy E_1, N_2 of them have energy E_2, etc.,

$$W = \frac{N!}{N_1!N_2!\ldots N_k!} \tag{4.2}$$

For large N, the Stirling transformation changes formula (4.2) into

$$\ln W = N \ln N - \sum_{i=1}^{k} N_i \ln N_i \tag{4.3}$$

4.3 Information Content of Structures

In 1949, Shannon [4] showed that the statistical concept of entropy can be extended beyond thermodynamics and applied to the process of transmitting information. Shannon's information theory regards a message transmitted through information channels as a specific set of symbols (an "outcome") selected from the ensemble of all *k* such sets containing the same total number of symbols *N*. Probabilities $p_1, p_2, \ldots, p_k$ are assigned to each of the outcomes, and the probability of the *i*th outcome is proportional to the number of symbols N_i it contains ($p_i = N_i / N$). Shannon's entropy of information *H* characterizes the uncertainty of the expected outcome. When the transmission is totally random all outcomes are equally probably and the entropy of information is the maximal one. Conversely, if there is a single possible outcome, H = 0. In the intermediate real-life cases the amount of information *I* transmitted is the difference between the maximum entropy and the specific value that Shannon's *H* function has for the system of interest. Thus, information has the meaning of reduced uncertainty of the final outcome.

The basic Shannon formula is obtained from equations (4.1) and (4.2), after taking the proportionality constant $k = 1 / \ln 2$ to measure the entropy of information in bits (binary digits).

$$H = N \log_2 N - \sum_{i=1}^{k} N_i \log_2 N_i, \text{ bits} \tag{4.4}$$

Another form of Shannon's equation determines the average entropy of information per communication symbol:

$$H_{av} = \frac{H}{N} = -\sum_{i=1}^{k} p_i \log_2 p_i = -\sum_{i=1}^{k} \frac{N_i}{N} \log_2 \frac{N_i}{N}, \text{ bits/symbol} \tag{4.5}$$

One bit of information, $H_{av} = 1$ bit, is obtained in learning the outcome of a situation in which there is a choice between two possible options.

One of the major consequences of Shannon's theory was the radically new idea of viewing a *structure* of any kind as a communication, which carries a certain amount of information. Thus, the notion of the *information content of a molecule* emerged in the early 1950s (Dancoff and Quastler [5], Linhitz [6], Rashevsky [7]), along with a generalization of the Second Law Thermodynamics that included information (Brillouin [8]). The Negentropy Principle of Information of Brillouin[8] regards information as a negative component of entropy. In this way, the generalized Second Law allows only such spontaneous processes in isolated systems that increase entropy or/and decrease information. Information cannot increase in irreversible processes; it can only diminish.

Mowshowitz [9] first presented in 1968 a rigorous reinterpretation of Shannon's H-function as information content but not entropy. He pointed out that Shannon's function does not measure the average uncertainty per structure of a given ensemble of all structures having the same number of elements, e.g., the selection of a molecule from the ensemble of all molecules having the same number of atoms. Rather, it is the information content of the structure relative to a system of symmetry transformations that leave the system invariant. In 1979, Bonchev [10] added the argument that entropy is transformed into information by the process of structure formation from its constituent elements. This makes the information a *bonded* one; it is conserved within the system until it is destroyed.

As could be anticipated from the preceding text, this chapter will not represent a general review or comparative analysis of complexity measures used in the realm of chemistry. Such an analysis is presented in the first several chapters of this volume. Here, we focus on the important question whether complexity measures based on Shannon's information theory could at all be adequate measures of *structural* complexity. Serious doubts about the positive answer of this question were raised in our previous publications [11–13]. It will be shown that the positive result is necessarily associated with a reformulation of the original information-theoretic formalism.

4.4 Symmetry-Based Information-Theoretic Indices

The information content of a structure, as defined in Section 4.3, is based on symmetry. The elements of the system are grouped into equivalence classes, according to a certain equivalence criterion or symmetry operation(s) that exchange the elements without violating the structure adjacency. Mowshowitz [9] formalized the application of Shannon's equations to finite systems with symmetry elements. He introduced a probability scheme applicable to any system having N elements partitioned into k classes according to the equivalence criterion α :

Equivalence classes	$1, 2, \ldots, k$
Element partition	$N_1, N_2, \ldots, N_k$
Probability distribution	$p_1, p_2, \ldots, p_k$

Here, $p_i = N_i / N$ is the probability for a randomly chosen element to belong to class i having N_i elements, and $\Sigma N_i = N$. Shannon's equations (4.4) and (4.5) can now be rewritten as equations for the information content ${}^eI(\alpha)$ of the system, and for the average information content ${}^eI_{av}(\alpha)$ of a system element. The left superscript e stands for "equivalence" to distinguish this type of information measure from those for "magnitude" introduced in Section 4.5.

$$ {}^eI(\alpha) = N \log_2 N - \sum_{i=1}^{k} N_i \log_2 N_i \tag{4.4a} $$

$$ {}^eI_{av}(\alpha) = \frac{{}^eI(\alpha)}{N} = -\sum_{i=1}^{k} p_i \log_2 p_i = -\sum_{i=1}^{k} \frac{N_i}{N} \log_2 \frac{N_i}{N} \tag{4.5a} $$

The values of both information indices thus introduced vary within the following ranges:

$$ 0 \leq {}^eI(\alpha) \leq N; \qquad 0 \leq {}^eI_{av}(\alpha) \leq 1 \tag{4.6} $$

The upper bound of these ranges is reached when each element forms a separate class, $N_i = 1$, i.e., when the system has no symmetry. The information content is zero when all of its elements are equivalent, $N_1 = N$, i.e., when the system has no structure, due to its very high symmetry. One might infer that the information indices ${}^eI(\alpha)$ and ${}^eI_{av}(\alpha)$ could be used as complexity measures, which relate higher complexity to asymmetry and a larger diversity of system elements. Low complexity (simplicity) is characterized as uniformity or lack of diversity, resulting from high symmetry.

4.4.1 Atomic Information Content

Interesting results are obtained when information theory is applied to atoms and molecules on a quantum-mechanical level. When electrons are regarded as distributed over spin orbitals, according to the Pauli exclusion principle each electron must occupy a different orbital. Thus, all N_i's in equations (4.4a) and (4.5a) are equal to 1, and the information content of the respective atom or molecule is the maximum one for the given number of electrons. The Pauli principle was reinterpreted as a principle for the maximum information content of atoms and molecules or, more generally, of fermionic systems [14]. Conversely, photons or, more generally, bosonic particles can populate the same quantum state in unlimited numbers. All particles in this kind of system are equivalent, and the information content is zero.

Other information-theoretic indices have been introduced for atoms and molecules to distinguish those of them having the same number of particles. *Atomic information content* has been introduced in various ways depending on the distribution of electrons, protons and neutrons in the electron shell and atomic nucleus [15,16]. As an example, consider the electron distribution into atomic orbitals (AOs) for the sulfur atom. According to the Pauli exclusion principle, only one or two electrons can populate an atomic orbital. The sixteen sulfur electrons occupy seven AOs with two electrons each, and two AOs with one electron each. Applying equation (4.4a), one obtains $I(S, AO) = 16 \log_2 16 - 7.2 \log_2 2 - 2.1 \log_2 1 = 66$ bits per atom, and $I_{av}(S) = 66/16 = 4.125$ bits per electron.

In assessing the complexity of atoms, information-theoretic indices have found application by providing a new systematic of nuclides [15], a reinterpretation of the Periodic Table of the chemical elements [17], and predictions of the properties of transactinide elements 113–118 [18]. A trend has been found which shows that the atomic information content increases the most when a new electron shell or s-, p-, d- and f-subshell begins. This made it possible to predict the atomic number of the first g-element [17].

4.4.2 Molecular Information Content

The information-theoretic formulas (4.4a) and (4.5a) have been used to characterize molecular complexity in various ways, depending on the structural elements taken as a basis, and the equivalence criterion used to group these elements into classes. The most straightforward way is to proceed from the chemical formula of the compound, which represents the elements incorporated as well as the ratio in which the atoms of the elements occur. Thus, Dancoff and Quastler [5] defined the information on the chemical composition $^{e}I(cc)$ of a molecule in 1953 by using the nature of the chemical element as an equivalence criterion. As an example, consider the molecule of potassium hydrogen phosphate, K_2HPO_4. The eight atoms of this molecule are distributed into four elemental classes, and the formulas (4.4a) and (4.5a) produce $^{e}I(cc) = 8 \log_2, 8 - 4 \log_2 4 -$

$2 \log_2 2 - 2.1 \log_2 1 = 14$ bits per molecule and ${}^{e}I_{av}(cc) = 14/8 = 1.75$ bits per atom. Information on the chemical composition of a molecule demonstrates the potential of the information content of a system to characterize its complexity regarded as the diversity of its elements. An illustration of this conclusion is the series of halogen derivatives of methane. ${}^{e}I(cc)$ increases systematically in the sequence of molecules CH_4, CH_3F, CH_2FCl, CHFClBr: 3.61, 6.86, 9.61, and 11.61 bits per molecule; respectively.

Unlike information on elemental composition, *structural information content* can be characterized in a variety of ways, proceeding from the molecular geometry or from the connectivity of atoms. Regarding atoms as the simplest structural elements of a molecule, one may use symmetry to group them into equivalence classes. When symmetry is characterized by the corresponding point group of the molecule one defines the *information index on molecular symmetry*,[19] ${}^{e}I_{sym}$.

Alternatively, when only atom-atom connectedness is taken into consideration but not specific molecular geometry, a molecule is represented by a molecular graph. The simplest criterion for equivalence of the graph vertices is their degree, i.e., the number of their nearest neighbors, and this structural information index is termed the *information on the vertex degrees*, ${}^{e}I(deg)$. Despite its simplicity this criterion is very relevant in chemistry where it corresponds to the coordination number in crystals and coordination compounds. It provides the classification of carbon atoms in organic chemistry as primary, secondary, tertiary, and quaternary ones.

When the equivalence criterion is extended over all atoms, the resulting vertex distribution determines the *topological information index* of Rashevsky [7]. The most precise definition of vertex equivalence was given by Trucco [20] and is based on the automorphism group of the graph. Equivalent graph vertices are those that belong to the same orbit of this group of symmetry. Due to the fact that many other topological information indices have been proposed after Rashevsky, we prefer to call this index the *information on the vertex orbits*, ${}^{e}I(orb)$ [21].

As an example for the above-mentioned types of structural information content, consider the molecule of 2,2,3,3-tetramethyl pentane and its graph (Figure 4.2). Let the molecule be taken in its most symmetric conformation belonging to the C_s group. Atoms 1–5 are positioned in the horizontal xy-plane, whereas the pairs of symmetric atoms 6,7 and 8,9 are in the vertical *xz*-plane. The set of nine carbon atoms are then distributed into two classes of two atoms and five classes of one atom each. Equations (4.4a) and (4.5a) produce for this case ${}^{e}I_{sym} = 24.53$ bits per molecule and 3.17 bits

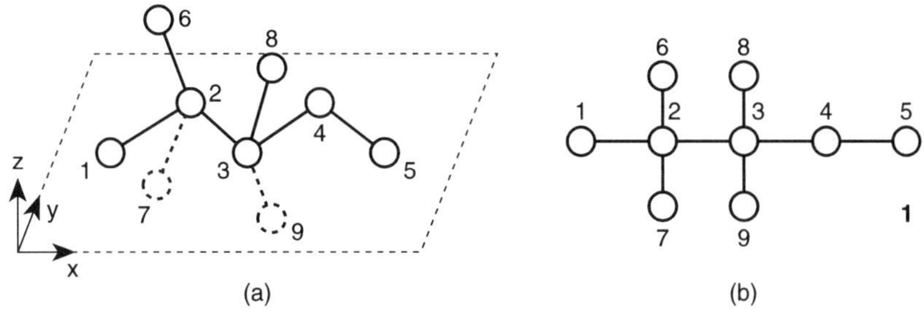

Figure 4.2: (a) The point group symmetry of the 2,2,3,3-tetramethylpentane molecule makes equivalent the pairs of atoms 6,7 and 8,9; (b) The automorphism group of the molecular graph makes equivalent vertices 1,6,7 and 8,9.

per atom. The distribution of the nine vertices in the molecular graph according to their vertex degrees is two vertices of degree four, one vertex of degree two, and six vertices of degree one. The resulting value of the information on the vertex degrees is ${}^{e}I(deg) = 11.02$ bits per molecule and ${}^{e}I_{av}(deg) = 1.22$ bits per atom. When graph vertices are grouped into the graph orbits, vertices 1, 6, and 7 belong to the same orbit. Another orbit includes vertices 8 and 9. The remaining four vertices are in an individual orbit each. The distribution 9(3,2,1,1,1,1) determines the total information on the graph orbits of this molecule to be ${}^{e}I(orb) = 21.77$ bits, and the average information is ${}^{e}I_{av}(orb) = 2.42$ bits per atom.

Up to here we have introduced three different criteria for determining equivalent atoms in a molecule by regarding atoms as the simplest structural elements. Graph edges can also be used as structural elements and the orbits of the corresponding edge group of the graph can determine their equivalence classes. Trucco mentioned this kind of information index [20]. We shall call it the *information on the edge orbits* of the graph, ${}^{e}I(edge\ orb)$. There are eight edges in the molecular graph **1** of 2,2,3,3-tetramethylpentane (Figure 4.2). According to their symmetry they form five edge orbits: {12, 26, 27), {38, 39), (23}, {34}, and {45}. The respective values of the information indices are ${}^{e}I(edge\ orb) = 8\log_2 8 - 3\log_2 3 - 2\log_2 2 = 17.25$ bits per molecule, and ${}^{e}I_{av}(edge\ orb) = 2.16$ bits per edge.

The next more complex type of substructure after vertices and edges are the subgraphs containing two adjacent edges. Gordon and Kennedy [22] were the first to use this number as a branching index. Even earlier, Platt [23] introduced a twice-larger index as the sum of the first bond neighbors of each bond in the molecule. Bertz defined this type of substructure more broadly as having a "pair of adjacent edges (connections)" and including multiple edges and loops [24]. One can also define the automorphism group of graph connections. Its orbits determine the *information on connections orbits*, ^eI *(conn orb)*. There are twelve connections in graph **1**, and they are distributed into the following orbits: {126, 127, 267}, {123, 623, 723}, {238, 239}, {438, 439}, {234}, and {345}. The distribution 12 {3,3,2,2,1,1} yields ^eI *(conn orb)* = 29.51 bits per molecule and $^eI_{av}$ *(conn orb)* = 2.46 bits per connection.

One might continue applying the Shannon equations to larger and larger substructures. Instead, we will inspect several more indices of this class based on different types of structural invariants. Two such indices have been introduced by proceeding from the equivalence of distances in a molecular graph, ^eI*(dist)* [25,26], and the equivalence of vertices being equidistant to the graph center, ^eI*(centric)* [27,28]. The distance between two vertices is the number of edges along the shortest path that connects them; therefore, all graph distances are integers. Consider again as an example graph **1** (Figure 4.2b). There are 36 distances in this graph: 8 distances of length 1, 13 distances of length 2, 12 distances of length 3, and 3 distances of length 4. The distance distribution 36{8, 13,12, 3} results in ^eI*(dist)* = 66.24 bits per molecule and $^eI_{av}$*(dist)* = 1.84 bits per distance.

The *centric information index*, ^eI*(centric)* [27,28] is based on the definition of the graph center [29]. The classical definition of Jordan, given in the nineteenth century, specifies the center of trees (acyclic graphs) with a procedure of consecutive pruning the tree by cutting off all of its terminal vertices. The final result is a single center (a vertex) or a bicenter (an edge). In 1969, Harary [29] proposed a rigorous definition that also includes cyclic graphs. It introduced the notion of the *vertex eccentricity* *e*(*i*) as the longest distance from a given vertex to any other vertex in the graph. The graph center is the vertex with the minimum eccentricity:

$$e(i) = \max d(ij) \quad ; \quad e(i) = \min \tag{4.7}$$

The pitfall of this definition is that it often produces a group of nonequivalent central vertices. Bonchev, Balaban and Mekenyan [27] proposed a more detailed solution in which the centric vertices belong to the same orbit of the automorphism group of the graph. The classical definition (4.7) is regarded only as a first of several hierarchically applied criteria, which gradually reduce the number of central vertices until only a single vertex or two or more equivalent vertices remain. The second criterion requires the central vertex to have the smallest sum of distances to all the remaining vertices in the graph. This sum is termed the *vertex distance*, $d(i) = \Sigma\ d(ij)$, or *distasum* [30]. If two or more nonequivalent vertices have the same minimal eccentricity and the same vertex distance then the third criterion, requiring a minimal occurrence of the largest distance, $n(i,j;\max)$, is used:

$$d(i) = \sum\nolimits_j d(i,j) = \min \tag{4.8}$$

$$n(i,j;\max) = \min \tag{4.9}$$

When the three criteria fail to produce equivalent centric vertices, an Iterative Vertex-Edge Centricity (IVEC) algorithm is applied [31], which always solves the problem with only one or two iterations.

In the example of graph **1** the first criterion suffices to determine vertex 3 as the graph center, due to its minimal eccentricity $e(3) = 2 = \min$. Vertices 2, 4, 8, and 9 have eccentricity 3, and vertices 1, 5, 6, and 7 have eccentricity 4. Let us now order the graph vertices centrically, according to their distance from the graph center. Two layers of equidistant vertices are thus formed around the center with vertices 2, 4, 8, and 9 in the first neighborhood layer, and vertices 1, 5, 6, and 7 in the second one. Thus, the nine vertices of this graph have a centric distribution 9 {1, 4, 4), from which ${}^{e}I(centric) = 12.53$ bits per molecule and ${}^{e}I_{av}(centric) = 1.39$ bits per atom.

Several information indices have been developed by proceeding from combined equivalence criteria. In fact, the first topological information index of Rashevsky [7] combines two criteria of atom equivalence: in order to be equivalent two atoms in a molecule should belong to the same chemical element and have the same atomic neighborhoods up to the terminal atoms. Basak et al. [32–34] used the same criteria but with separate terms for the first, second, …, *k* th neighborhood. Basak's *k*th order *neighborhood complexity index* in hydrogen-suppressed graphs is thus identical to Rashevsky's index, and for such graphs having no heteroatoms his first order index coincides with the information on vertex degrees ${}^{e}I(deg)$. Basak applied his neighborhood indices not only to hydrogen-suppressed molecular graphs but also to molecular graphs that include hydrogen atoms and heteroatoms, and his methodology was used with success in a wide range of QSAR studies and assessments of environmental toxicities [34].

Other information indices based on a combination of structural criteria have also been introduced [10,21,26,35]. Like the indices described above

(Table 4.1, *vide infra*), they reflect only some of the complexity features but fail for others. For this reason, we add here only the *molecular complexity* index of Bertz [24, 36], *BI*, which satisfies most of the requirements for a complexity measure. Bertz's index combines the information on graph connections with a size term $N \log_2 N$, with N being the total number of connections:

$$BI = 2N \log_2 N - \sum_{i=1}^{k} N_i \log N_i \tag{4.10}$$

Adding the size term reduces the inherent pitfalls of symmetry-based information complexity measures. The corrective term allows us to obtain different nonzero values for highly symmetrical molecules like the cycloalkanes. It reflects to a certain extent molecular complexity features like branching and cyclicity. Thus, it increases the structural component of the information measure and diminishes the simplifying effect of symmetry.

It is essential to verify the extent to which the information-theoretic indices introduced so far, proceeding from the differently defined equivalence classes of structural invariants, satisfy the intuitive idea of complexity. Experts generally agree that path (or linear) graphs P_n are the simplest connected graphs and the complete graphs K_n are the most complex ones. There is also general agreement that for the same number of vertices the star graphs S_n and monocyclic graphs C_n are more complex than path graphs and less complex than complete graphs. Bertz and Zamfirescu [37] formalized these criteria in the form of inequalities which a complexity measure must satisfy.

We extend this series of four graphs with two more, the first being totally disconnected, and the second being bicyclic. Indeed, the intuitive expectation is that the disconnected graph will be less complex than any connected graph including P_n. Similarly, the bicyclic graph should be more complex than the monocyclic one. Figure 4.3 shows the selected six graphs each having five vertices. The values of the calculated information indices

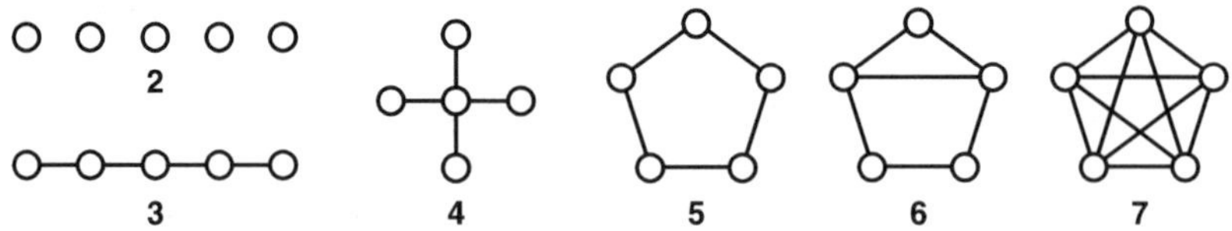

Figure 4.3: Graphs with five vertices used as an illustration of increasing structural complexity.

Table 4.1: Information indices based on the equivalence of vertices, edges, connections, centrically ordered vertices, and distances cannot reproduce the increasing structural complexity of the sequence of the five-vertex graphs **2–7**

Index	**Graphs**					
	2	**3**	**4**	**5**	**6**	**7**
^{e}I(vertex, *deg*)	0	7.61	3.61	0	4.86	0
^{e}I(*vertex*)	0	7.61	3.61	0	7.61	0
^{e}I(*edge*)	0	4	0	0	11.51	0
^{e}I(*connection*)	0	2.75	0	0	20.53	0
^{e}I(*vertex, centric*)	0	7.61	3.61	0	7.61	0
^{e}I(*distance*)	0	18.47	9.71	10	9.71	0
BI (Bertz index)	0	2.75	15.51	11.61	49.06	147.21

are shown in Table 4.1. Since the first four indices are calculated from the orbits of the respective automorphism groups, for brevity we omit the abbreviation "orb", and denote only the type of structural element used: vertices, edges, connections, and centrically ordered vertices. The Basak set of indices [32–34] were not included because for hydrogen-suppressed graphs that have no heteroatoms they include *I*(*deg*) and *I*(*vertex orbits*).

Although the six graphs have the same number of vertices, the number of edges increases from zero in **2**; to four in **3** and **4**, to five in **5**, to six in **6**, and to ten in **7**. The number of connections increases even faster: from zero in **2**, to three in **3**, to six in **4**, to five in **5**, to nine in **6**, and to twenty one in **7**. Intuition implies that the structure is more complex when the number of the interconnections of its elements increases. However, the first five indices shown in Table 4.1 fail to distinguish between the disconnected graph **2**, the monocycle **5**, and the complete graph **7**; ascribing zero information content to all of them. The information on the equivalence of distances decreases with increasing branching and cyclicity. This is well reflected by the values of star versus path graph, as well as in the sequence monocyclic $\Rightarrow$ bicyclic $\Rightarrow$ complete graph. However, the equal complexity of disconnected and complete graphs is a major failure of this index. Another problem of most equivalence-based indices is their opposing complexity trends to decrease with branching and cyclicity and increase with size.

One might conclude that replacement of the vertices, which are the simplest subgraphs, with larger subgraphs containing two and respectively three vertices, as well as the use of graph distances will not produce an adequate complexity measure. The real meaning of these symmetry-based indices is that they measure diversity. Diversity coincides with complexity when dealing with the chemical composition of molecules. However, as manifested by the examples in Table 4.1, structural diversity and structural complexity have little in common. Yet, the last line in Table 4.1 shows that with a conveniently selected corrective term, the equivalence of some more sensitive graph invariants, like the two-edge subgraphs used by Bertz, could provide better complexity measures. Yet, as can be inferred from the more detailed evaluation [38], the Bertz's index [24] faced difficulties when dealing with more subtle structural patterns, and the search for new solutions continued.

4.5 Magnitude-Based Information Measures of Complexity

An extension of the Shannon information-theoretic approach to the description of chemical structures was reported in 1977 by Bonchev and Trinajstić [25]. The finite probability scheme of Mowshowitz [9] was expanded so as to include certain weights or magnitudes of the structure elements. The scheme can be applied to any system having N elements partitioned into k classes according to the element weight (or magnitude) of type α :

Equivalence classes	$1, 2, \ldots, k$
Element partition	$N_1, N_2, \ldots, N_k$
Probability distribution	$p_1, p_2, \ldots, p_k$
Magnitudes (weights)	$w_1, w_2, \ldots, w_k$
Probability M-distribution	${}^m p_1, {}^m p_2, \ldots, {}^m p_k$

Here, $\Sigma\ N_i w_i = M$, where M is the magnitude of the criterion (or property) selected to partition the system elements, $p_i = w_i /M$ is the probability for a randomly chosen element to belong to class i having magnitude w_i and $\Sigma\ {}^m p_i = 1$. Shannon's equations (4.4) and (4.5) define in this approach the magnitude-based information content ${}^m I(\alpha)$ of the system, and the corresponding average information content ${}^m I_{av}(\alpha)$ per system element:

$$ {}^{m}I(\alpha) = M\log_2 M - \sum_{i=1}^{k} N_i w_i \log_2 w_i \tag{4.11} $$

$$ {}^{m}I_{av}(\alpha) = -\sum_{i=1}^{k} N_i \frac{w_i}{M}\log_2 \frac{w_i}{M} \tag{4.12} $$

Information measures (4.11) and (4.12) are defined within the ranges

$$ 0 \leq {}^{m}I(\alpha) \leq M\log_2 M : 0 \leq {}^{m}I_{av}(\alpha) \leq \log_2 M \tag{4.13} $$

where the lower bound corresponds to a system without a structure ($w_i = M$), and the upper bound can be attained by a system having a maximum number of classes with a single element of unit weight in each class ($k = N = M$).

The magnitude-based information indices were first introduced by Bonchev and Trinajstić [25] for the distribution of distances in a molecular graph, ${}^{m}I(distance)$ or ${}^{m}I_d$. The total magnitude characterizing the structure here is the *Wiener number* [39] W, which is the sum of all graph distances (the total graph distance). The classes of distance are those of distances $d(i) = 1, 2, 3, \ldots, d(\max)$, and the number of distances in these classes is $N_1, N_2, \ldots, N_k$, respectively, Thus the general formulas (4.11) and (4.12) are transformed into

$$ {}^{m}I(dist) = W\log_2 W - \sum_{i=1}^{k} N_i d(i) \log_2 d(i) \tag{4.14} $$

$$ {}^{m}I_{av}(dist) = -\sum_{i=1}^{k} N_i \frac{d(i)}{W}\log_2 \frac{d(i)}{W} \tag{4.15} $$

A second magnitude-based information measure is defined from the decomposition of the sum of distances in the graph distance matrix (doubled Wiener number) into contributions of different vertices, d_i. Termed the *distance degree* of vertex i (or *distasums*), these quantities represent the sum of the distances between the vertex and the remaining graph vertices. Thus, the vertex decomposition of the total graph distance is presented by the formulas:

$$ {}^{m}I(dist\ \deg) = 2W\log_2 2W - \sum_{i=1}^{N} d_i \log_2 d_i \tag{4.16} $$

$$^{m}I_{av}(dist\ \deg) = -\sum_{i=1}^{N} \frac{d_i}{2W} \log_2 \frac{d_i}{2W} \tag{4.17}$$

Graph vertices can be partitioned into magnitude-based classes in various ways. Besides the distance degrees distribution introduced in the previous paragraph, vertex distributions based on other criteria (or "degrees") can be used. In fact, each symmetry-based distribution can be transformed into a magnitude-based one. The first case of vertex distribution discussed in Section 4.4 dealt with the equivalence of vertex degrees, as inferred from the adjacency matrix of the graph. The sum of the vertex degrees, a_i, $A = \Sigma\ a_i$, is called the graph *total adjacency* [35]. This is the simplest magnitude-type graph invariant that can be used to construct an information descriptor. The vertex degree distribution $A\ \{a_1, a_2, \ldots, a_k\}$ produces the information index on vertex degree magnitudes [35], $^{m}I(vert\ deg)$:

$$^{m}I(vert\ \deg) = A \log_2 A - \sum_{i=1}^{N} a_i \log_2 a_i \tag{4.18}$$

$$^{m}I_{av}(vert\ \deg) = -\sum_{i=1}^{N} \frac{a_i}{A} \log_2 \frac{a_i}{A} \tag{4.19}$$

The graph connections (subgraphs having three connected vertices) used as the basis of the Bertz [24] index *BI*, can also be used to construct a vertex distribution $C\{C_1, C_2, \ldots C_n\}$, according to the number of connections C_i beginning in each vertex. Here C is the doubled number of graph connections. The magnitude-based information index on the graph connections, ^{m}I(conn deg), we propose here is thus defined as

$$^{m}I(conn\ \deg) = C \log_2 C - \sum_{i=1}^{N} C_i \log_2 C_i \tag{4.20}$$

$$^{m}I_{av}(conn\ \deg) = -\sum_{i=1}^{N} \frac{C_i}{C} \log_2 \frac{C_i}{C} \tag{4.21}$$

In Table 4.2 we analyze the capability of the four magnitude-based information indices, discussed in this section, to be used as complexity measures. Several conclusions can be drawn. It is seen that the total information indices show a considerably more consistent complexity trend than the average ones. The two distance-based measures demonstrate a

Table 4.2: Values of four magnitude-based information indices for graphs **2–7**

Index	**Graphs**					
	2	**3**	**4**	**5**	**6**	**7**
$^{m}I(distance)$	0	64.93	52.00	48.60	45.30	33.22
$^{m}I_{av}(distance)$	0	3.245	3.250	3.240	3.236	3.322
$^{m}I(dist\ deg)$	0	91.63	73.39	69.66	64.86	46.44
$^{m}I_{av}(dist\ deg)$	0	2.298	2.293	2.322	2.316	2.322
$^{m}I(vert\ deg)$	0	18.00	16.00	23.22	27.51	46.44
$^{m}I_{av}(vert\ deg)$	0	2.250	2.000	2.322	2.293	2.322
$^{m}I(conn\ deg)$	0	13.51	24.00	23.22	64.60	139.32
$^{m}I_{av}(conn\ deg)$	0	2.250	2.000	2.322	2.936	2.322

very regular pattern of *decreasing* with the increase in graph complexity. Therefore, these two indices are convenient complexity measures for isomeric molecules. The opposing trend to *increase* with the system size, however, prevents their use as complexity estimates of structures having different size. Nevertheless, these indices have a good potential for QSPR/QSAR applications when used jointly with a convenient size term. The index on the vertex degrees distribution increases with both size and cyclicity (cyclic complexity) [40, 41] but fails to show the same trend with branching (acyclic complexity) [25, 42–45]. Yet, the correct trend shown in cyclic molecules indicates that the magnitude-based information indices could in

8
20.00

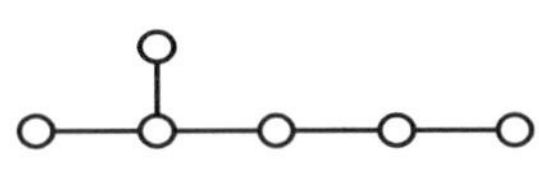

9
24.46

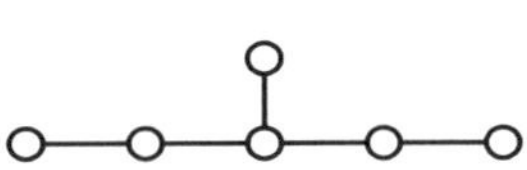

10
25.22

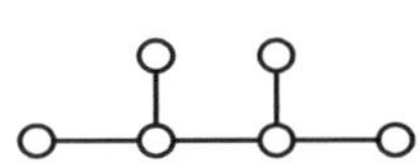

11
31.02

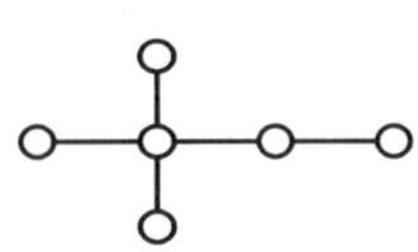

12
34.28

Figure 4.4: Graphs of the five acyclic hexane molecules ordered according to their increasing complexity, as assessed by the magnitude-based information index on the vertex connection degrees, ^{m}I (conn deg).

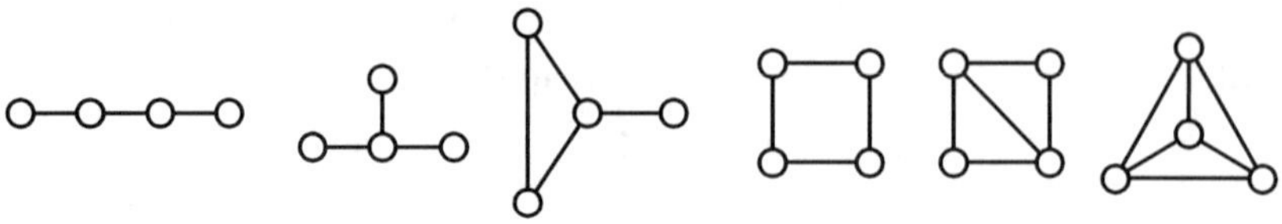

13	14	15	16	17	18
8	9.51	15.71	16	32	48

Figure 4.5: All four-vertex graphs ordered according to their increasng complexity.

principle provide a good complexity measure if a more detailed structural invariant was taken as a basis.

We investigated this idea by introducing the new ^{m}I (conn deg) index, based on the distribution of graph connections over the vertices from which they emanate. As seen in Table 4.2, the idea is a promising one, the new index showing a very regular increase with the increase in both branching and cyclicity. It also increases with the size of the system as another complexity component. Thus, its values for C3–C6 normal alkanes are 2, 8, 13.51, and 20 bits per molecule, and for C3–C6 cycloalkanes they are 9.51, 16, 23.22, and 31.02, respectively. The sensitivity toward subtler branching patterns can be illustrated with the series of five isomeric acyclic hexanes **8–12** (Figure 4.4). The new information index increases strongly with factors such as the number of branches and the degree of the branched vertex. A moderate increase is found when the branch is shifted toward a more central position.

A similar, systematic increase in the ^{m}I (conn deg) index values with the increase in branching and cyclicity can be seen in Figure 4.5 for all graphs having four vertices. Graphs **13–18** in Figure 4.5 are identical with graphs **VI–XI** in Chapter 2, used there as a basis for comparison of a large set of complexity measures. The ordering provided by our new index: **VI** $\Rightarrow$ **VII** $\Rightarrow$ **VIII** $\Rightarrow$ **IX** $\Rightarrow$ **X** $\Rightarrow$ **XI** does not coincide with any of the orderings analyzed in that chapter. It is close to the ordering provided by the Minoli index [47] and the number of spanning (or maximal) trees [3, 48–51], which, however, cannot distinguish between isomeric trees (acyclic graphs).

Thus, the ^{m}I (conn deg) index mirrors the complexity trends in both acyclic and cyclic molecules by proceeding from a general theoretical scheme but without adding a corrective term to the symmetry-based index, as is done in Bertz's treatment of graph connections distributions. More details on the new index will be given in a forthcoming publication [46]. We may conclude that information-theoretic indices based on magnitude distributions provide a better basis for the complexity assessments of molecules than the symmetry-based measures. However, only those magnitude-based distributions of graph invariants that describe the structure in more detail could be good measures of structural complexity. Graph connections are only the lowest level of a more complete description of a structure. One could easily infer that in graphs whose vertices all have the same degree (e.g., the graphs of the cubane molecule, $(CH)_8$ and its isomers) one would need to up one level and deal with distributions of three-edge substructures. One might thus also suggest that it is not the specific mathematical function used (Shannon's entropy function) but rather the detailed description of the structure that enables the construction of more reliable complexity measures. Developments along these lines are discussed in the next section.

4.6 Substructure-Based Information Measures of Complexity

A breakthrough in the methods of assessing the complexity of structures has been achieved during the last five years. The novel concept of complexity may be summarized by the sentence: "The more substructures in a system, the more complex the system." This concept is in agreement with our intuitive understanding that while the size of a system contributes to its complexity, it is the connectedness of the system elements that matters more. A larger system with weak interrelations of its elements may be regarded less complex that a smaller system with a high degree of internal connectedness. The more connected the system, the higher the number of substructures K in it. Then, why not simply count how many substructure there are in a structure?

Bertz and Herndon [52] briefly mentioned such an idea in 1986 in work devoted to measures of molecular similarity. In 1996–1998, simultaneously and independently, Bertz [53, 54] and Bonchev [55–58] developed this complexity concept in detail. Bone and Villar [59] used a similar approach to characterize molecular diversity in detail. Bertz also proposed to use the number of kinds of subgraph N_s as a measure of structural diversity and applied his complexity measures to determine the "strategic" bond, the creation of which in a synthetic reaction would increase the complexity of the molecule the most. Bonchev developed a substructure-based complexity concept by constructing a *complexity vector K′*

$$K'\{{}^0K, {}^1K, {}^2K, \ldots, {}^EK\}; \qquad K(G) = \sum_{e=0}^{E} {}^eK \tag{4.22}$$

This is an ordered sequence, the *e*th *order complexity* term, ${}^eK$, in which counts the subgraphs having *e* edges; *E* in equation (4.22) is the total number of graph edges.

In parallel with the idea of using all substructures, Rücker and Rücker [60] proposed to use *all walks* in the graph. They have shown that *the total walk count, twc,* is a measure of graph complexity that mirrors the basic patterns of acyclic and cyclic complexity to a very high degree [38].

The substructure approach to complexity was developed further by the present author into the concept of *overall complexity indices* [61–63]. The idea is to ascribe a certain weight to each subgraph, select simple graph invariants as weights, and then to find the overall value of this index for the entire structure by summing up over all subgraphs. Proceeding from vertex degrees and vertex-vertex distances, the graph invariants selected for each subgraph *i* were the total adjacency A_i (the sum of the vertex degrees a_i) the total distance W_i (the Wiener number), and the first and second Zagreb indices [64, 65], $M1_i$ and $M2_i$ (the sum of all the squared vertex degrees Σa_i^2 and the sum of the products of the vertex degrees, $\Sigma a_i a_j$, over all edges $\{ij\}$, respectively). The complexity indices thus defined were termed the *overall connectivity* [58, 61], OC, the *overall Wiener number* [62], *OW* and the *overall Zagreb indices* [63], *OM*1 and *OM*2, respectively. The overall connectivity was defined in two versions *TC*l and *TC* (abbreviation for *topological complexity*). The vertex degrees in *TC* are taken from the molecular graph *G*, whereas in *TC*l they are taken from the corresponding subgraphs G_i. All overall indices *OI*(*G*) are also presented in vector form *OI*′(*G*) incorporating all *e*th order terms ${}^eOI(G)$ where $I = C, W, M1$ or $M2$ indicates the types of the overall index. The current value *e* of the number of edges in the subgraph runs from zero for zero-order complexity (complexity of vertices), to one for first-order complexity (complexity of edges), etc., to $e = E$ for the entire graph having *E* edges.

$$TC'(G) = TC\{{}^0TC, {}^1TC, {}^2TC, \ldots, {}^ETC\};\quad TC(G) = \sum_{e=0}^{E} {}^eTC(G) = \sum_{i=1}^{K} A_i(G_i \subset G) \tag{4.23}$$

$$OW'(G) = OW\{{}^0OW, {}^1OW, {}^2OW, \ldots, {}^EOW\}; OW(G) = \sum_{e=1}^{E} {}^eOW(G) = \sum_{i=1}^{K} W_i(G_i \subset G) \tag{4.24}$$

$$OM'(G) = OM\{{}^0OM, {}^1OM, {}^2OM, \ldots, {}^EOM\}; OM(G) = \sum_{e=1}^{E} {}^eOM(G) = \sum_{i=1}^{K} M_i(G_i \subset G) \tag{4.25}$$

In equations (4.23) and (4.25) we omit for brevity additional symbols used to distinguish between the two overall connectivities TC and $TC1$, and between the first and second Zagreb index, $M1$ and $M2$, respectively.

All of the three types of overall topological indices satisfy the requirement for a complexity measure. They increase with both size and connectedness of the structure, and are very sensitive toward subtle complexity patterns in acyclic and cyclic structures. They increase with the number and size of the branches and cycles, as well as with their more central location (a topological feature called *centrality*), and with the closer branch/branch or cycle/cycle location (features called *branch adjacency* and *cycle adjacency*, respectively). Some of these complexity patterns are illustrated in Figures 4.4 and 4.5 (vide infra), along with the information indices on substructure distributions, which we will define below.

The partitioning of the overall topological indices into their substructural components (equations 4.23–4.25) gives rise to a typical situation for applying our magnitude-based information-theoretic approach. Each of the three types of overall indices may thus be supplemented by an *overall information index*, defined on the set of eth-order terms. Due to the detailed description of the topological structure provided by these parent sets, one might anticipate these to be the best structural complexity indices that a Shannon-type function could produce. The *overall connectivity information indices*, ${}^{m}I(TC)$ and ${}^{m}I(TC1)$, the *overall Wiener information index*, ${}^{m}I(OW)$, and the overall *Zagreb information indices*, ${}^{m}I(OM1)$ and ${}^{m}I(OM2)$ are defined below along with the *substructure count information index*, ${}^{m}I(K)$:

$$ {}^{m}I(K) = K\log_2 K - \sum_{e=0}^{E} {}^{e}K\log_2 {}^{e}K; \quad {}^{m}I_{av}(K) = {}^{m}I(K)/K \tag{4.26} $$

$$ {}^{m}I(TC) = TC\log_2 TC - \sum_{e=0}^{E} {}^{e}TC\log_2 {}^{e}TC; \quad {}^{m}I_{av}(TC) = {}^{m}I(TC)/TC \tag{4.27} $$

$$ {}^{m}I(TC1) = TC1\log_2 TC1 - \sum_{e=1}^{E} {}^{e}TC1\log_2 {}^{e}TC1; \quad {}^{m}I_{av}(TC1) = {}^{m}I(TC1)/TC1 \tag{4.28} $$

The summation in equation (4.28) does not start from $e = 0$, because the vertex degrees in $TC1$ are those in the corresponding subgraphs, and for isolated vertices $a_i = 0$. Similarly, ${}^{0}OW = 0$ in equation (4.29) below because the distances between the isolated vertices are zero. The vertex degrees in the overall Zagreb information indices (equations 4.30 and 4.31) are those from the entire graph, and the second overall Zagreb index has a zero-

Table 4.3: Overall topological and overall information indices of the graphs from Figure 4.6

Indices	Graphs				
	3	4	5	6	19
K	15	17	20	54	57
$^{m}I(K)$	32.24	43.32	64.16	138.20	144.72
$TC1$	40	64	110	310	326
$^{m}I(TC1)$	78.84	115.92	232.99	686.34	721.67
TC	60	100	160	482	522
$^{m}I(TC)$	135.81	209.22	232.20	1136.39	1225.71
OW	56	80	190	510	483
$^{m}I(OW)$	101.32	137.60	335.04	994.51	975.92
$OM1$	110	292	320	1228	1407
$^{m}I(OM1)$	247.71	600.23	752.71	2897.50	3301.05
$OM2$	60	68	342	2015	2025
$^{m}I(OM2)$	135.39	143.91	709.11	4131.10	4409.24

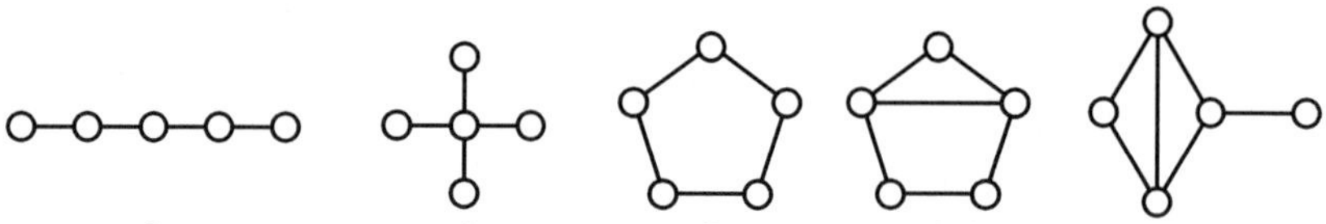

3 4 5 6 19

Figure 4.6: Five-vertex acyclic and cyclic graphs ordered according to their increasing complexity.

order term introduced by analogy with the zero-order term of the overall connectivity index TC as the sum of all vertex degrees in the graph.

$$^{m}I(OW) = OW\log_2 OW - \sum_{e=1}^{E} {}^{e}OW\log_2 {}^{e}OW; \quad {}^{m}I_{av}(OW) = {}^{m}I(OW)/OW \tag{4.29}$$

$$^{m}I(OM1) = OM1\log_2 OM1 - \sum_{e=0}^{E} {}^{e}OM1\log_2 {}^{e}OM1; \quad {}^{m}I_{av}(OM1) = {}^{m}I(OM1)/OM1 \tag{4.30}$$

$$^{m}I(OM2) = OM2\log_2 OM2 - \sum_{e=0}^{E} {}^{e}OM2\log_2 {}^{e}OM2; \quad {}^{m}I_{av}(OM2) = {}^{m}I(OM2)/OM2 \tag{4.31}$$

Table 4.3 summarizes the values of the overall topological indices *K*, *TC*, *TC*1, *OW*, *OM*1, and *OM*2 for graphs **3–6** and **19** from Figure 4.6. This is a modified Figure 4.3 with the totally disconnected graph **2** no longer shown because of its zero values for all indices. The complete graph **7** as the most complex graph having five vertices is characterized with very high values of all substructure-based indices, and is also not shown in Figure 4.6. Instead, the complexity of a new five-vertex graph **19** is analyzed with the expectation that it will be more complex than the other bicyclic graph **6**, because it has a branch as an additional complexity factor.

The expected ordering of these five graphs with their complexity increasing from the linear graph **3** to the star graph **4** to the monocyclic graph **5** to the bicyclic graph **6** to the bicyclic branched graph **19** is:

$$\mathbf{3} \Rightarrow \mathbf{4} \Rightarrow \mathbf{5} \Rightarrow \mathbf{6} \Rightarrow \mathbf{19}$$

As seen from Table 4.3, this ordering is confirmed by all indices examined. The only exception is found for the overall Wiener index and its information counterpart, which reorder graphs **6** and **19**. The bicyclic branched graph **19** has three subgraphs more than the unbranched bicyclic graph **6**. However, these are small size subgraphs having two and three edges per subgraph, and the increase in the corresponding *OW* terms they provide does not suffice to compensate for the larger number of longer linear subgraphs with four and five edges in graph **6**. Indeed, there is a competition between the two complexity factors of branching and cycle size in these two graphs. The overall Wiener index and the overall Wiener information indices appear to be the only ones to favor branching over cycle size.

A more detailed analysis of acyclic branching is made by comparing the complexity of the five isomeric acyclic hexanes (Figure 4.4 and Table 4.4). The magnitude-based overall information indices, as well as their parent overall topological indices, reflect the regular increase in complexity in the sequence

$$\mathbf{8} \Rightarrow \mathbf{9} \Rightarrow \mathbf{10} \Rightarrow \mathbf{11} \Rightarrow \mathbf{12}$$

Several complexity factors are thus correctly reflected: the increase in the number of branches (**8** $\Rightarrow$ **9,10** $\Rightarrow$ **11,12**), the branching at a vertex of higher degree (**11** $\Rightarrow$ **12**), and the shifting of a branch toward a more central position (**9** $\Rightarrow$ **10**). Notably, the change in the overall information indices upon the action of centrality factor (the **9** $\Rightarrow$ **10** transformation) is

Table 4.4: Overall topological and overall information indices of the five acyclic graphs **8–12** (Figure 4.4)

Indices	Graphs				
	8	**9**	**10**	**11**	**12**
K	21	24	25	28	30
$^{m}I(K)$	50.36	58.56	61.0	67.48	72.78
$TC1$	70	88	94	112	122
$^{m}I(TC1)$	159.55	195.47	206.01	239.83	258.03
TC	100	127	136	164	181
$^{m}I(TC)$	251.61	311.84	329.63	388.48	422.82
OW	126	154	161	188	197
$^{m}I(OW)$	263.92	315.14	328.32	373.74	392.74
$OM1$	188	277	300	404	505
$^{m}I(OM1)$	471.24	675.13	720.40	947.92	1165.91
$OM2$	130	149	161	172	168
$^{m}I(OM2)$	321.65	358.18	383.02	399.97	389.94

larger than that in the corresponding parent overall topological indices. This illustrates the high sensitivity of the newly introduced information indices toward subtle complexity patterns.

The ordering of the five hexanes coincides with the ordering produced by other symmetry-independent indices, whereas not a single symmetry-based index (including their best representative, the Bertz index [24]) is able to mirror all of the three branching patterns identified in this series of graphs [38]. The only partial reordering that is produced by the twelve overall topological and information indices deals with graphs **11** and **12**. The second overall Zagreb index $OM2$ and its information analogue $^{m}I(OM2)$ fail to show graph **12** as more complex than graph **11**. The reason for this result can be traced to the smaller index increment produced by the $OM2$ function for two neighboring vertices of degree four and two, as compared to two neighbors of degree three ($3{\times}3{>}4{\times}2$). Thus, the $OM2$ and $^{m}I(OM2)$ indices cannot always reflect the greater contribution to molecular complexity coming from branching at an atom of higher valence.

4.7 Concluding Remarks

This study revises the earlier criticism of the present author [11–13, 55] toward the use of the Shannon information function as a complexity measure. The findings in our previous studies were that the information content of a molecule, calculated by the standard symmetry- or equivalence-based scheme, is a good measure for molecular diversity, particularly for the atomic composition of molecules. At the same time, it was concluded that no information-theoretic index constructed on such a basis could measure the structural complexity of a molecule or any other system. The only complexity index of this type with relatively good performance is the Bertz index [24] based on graph connections. However, as shown in this chapter, the parent symmetry-based information index on the graph connections is a very poor complexity measure. The success of the Bertz index is due to the very essential addition of a "size term," which in most cases compensates for the weaknesses inherent for the information-theoretic technique employed. The correction incorporated is not based on a rigorous theory and has only practical importance. Yet, even this index faces difficulties in reflecting some subtle complexity patterns in acyclic and cyclic systems.

The second information-theoretic scheme, developed by Bonchev and Trinajstić [25] in 1977, is based on the partitioning of certain magnitudes or weights of the system into the contributions of its elements, and is thus not related to the symmetry of the system. The magnitude-based information indices on graph distances, ${}^{m}I$(*distance*), introduced in that study mirror to a high extent molecular complexity patterns in *isomeric* acyclic and cyclic systems. When combined with another index to account for the size of the system, this index has found application in structure-property and structure-activity relationships (QSPR, QSAR) studies [66, 67]. However, the ${}^{m}I$(*distance*) index does not yield a good complexity measure for systems of different size, owing to the opposing trends of increasing with size and decreasing with branching and cyclicity. For this reason, the conclusion was drawn in our previous work [11–13, 55] that the Shannon information function could not be a measure of structural complexity.

In the present study, these rumors of the premature death of information-theoretic complexity measures were found to be strongly exaggerated, though they have gained further ground specifically for the *symmetry-based* information indices. The revision refers to the *magnitude-based* information complexity measures, provided the graph invariant used to construct the magnitude distribution is selected so as *to increase* with both size and structural complexity factors. We have proposed in this chapter several such indices, beginning with the ${}^{m}I$(conn deg) index. It is based on the distribution of graph vertices according to the number of connections (two-edge subgraphs) that emanate from each vertex. Indeed, the number of graph connections is not a universal graph invariant, and cannot serve as a basis for a universal information-theoretic complexity measure. It cannot, for example, distinguish graphs in which all vertices belong to a single orbit. This indicated that the distributions of more complex subgraphs having three or more edges should be addressed as well. Instead of these potentially useful but only partial solutions to the problem, we turned to the complete structural characterization of a graph by the magnitude-type distribution of all subgraphs in the graph, as well as by a similar distribution provided by the recently developed overall topological indices [58, 61–63]. The latter ascribe certain weights or magnitudes to each of the subgraphs, such as the subgraph total adjacency, [58, 61] its total distance, [62] and the vertex-degree-based Zagreb indices [63]. The overall information indices thus constructed and, particularly, the two information overall connectivity indices and that on the total substructure count would seem to satisfy to a high degree the requirements for a good measure of structural complexity.

4.8 References

1. O.N. Temkin, A.V. Zeigarnik, and D. Bonchev, *Chemical Reaction Networks. A Graph Theoretical Approach*, CRC Press, Boca Raton, FL, 1996.
2. O.N. Temkin, A.V. Zeigarnik, and D. Bonchev, Graph-Theoretical Models of Complex Reaction Mechanisms and Their Elementary Steps, in: *Graph-Theoretical Approaches to Chemical Reactivity*, D. Bonchev and O. Mekenyan (Eds), Kluwer Academic, Dordrecht, 1994, pp. 143–278.
3. K. Gordeeva, D. Bonchev, D. Kamenski, and O.N. Temkin, Enumeration, Coding, and Complexity of Linear Reaction Mechanisms, *J. Chem. Inf. Comput. Sci.*, **34**, 244–247 (1994).
4. C. Shannon and W. Weaver, *Mathematical Theory of Communication*, University of Illinois Press, Urbana, IL, 1949.
5. S.M. Dancoff, and H. Quastler, The Information Content and Error Rate of Living Things, in: *Essays on the Use of Information Theory in Biology*, H. Quastler (Ed.), University of Illinois Press, Urbana, IL, 1953.
6. H. Linshitz. The Information Content of a Bacterial Cell. In: *Essays on the Use of Information Theory in Biology*, H. Quastler (Ed.), University of Illinois Press, Urbana, IL, 1953.

7. N. Rashevsky, Life, Information Theory, and Topology. *Bull. Math. Biophys.*, **17**, 229–235 (1955).
8. L. Brillouin, *Science and Information Theory*, Academic Press, New York, 1956.
9. A. Mowshowitz, Entropy and the Complexity of Graphs: I. An Index of the Relative Complexity of a Graph, *Bull. Math. Biophys.*, **30**, 175–204 (1968).
10. D. Bonchev, Information Indices for Atoms and Molecules. *MATCH*, **7**, 65–113 (1979).
11. D. Bonchev and D.E. Polansky, On the Topological Complexity of Chemical Systems, in: *Graph Theory and Topology in Chemistry*, R.B. King and D.H. Rouvray (Eds), Elsevier, Amsterdam, 1987, pp. 126–158.
12. D. Bonchev, The Problems of Computing Molecular Complexity, in: *Computational Chemical Graph Theory*, D.H. Rouvray (Ed.), Nova Publications, New York, 1990, pp. 34–67.
13. D. Bonchev and W.A. Seitz, The Concept of Complexity in Chemistry, in: *Concepts in Chemistry: A Contemporary Challenge*, D.H. Rouvray (Ed.), Research Studies Press, Taunton, UK, 1996, pp. 348–376.
14. D. Bonchev, Information Theory Interpretation of the Pauli Principle and Hund Rule, *Intern. J. Quantum Chem.*, **19**, 673–679 (1981).
15. B. Rousseva and D. Bonchev, A Theoretic-Information Variant of Nuclide Systematics. *Commun. Math. Chem* (*MATCH*), **4**, 173–192 (1978).
16. D. Bonchev, V. Kamenska, D. Kamenski, Informationsgehalt Chemischer Elemente, *Monatsh. Chem.*, **108**, 477–487 (1977).
17. D. Bonchev and V. Kamenska, Information Theory in Describing the Electronic Structure of Atoms, *Croat. Chem. Acta*, **51**, 19–27 (1978).
18. D. Bonchev and V. Kamenska, Predicting the Properties of the 113–120 Transactinide Elements, *J. Phys. Chem.*, **85**, 1177–1186 (1981).
19. D. Bonchev, D. Kamenski, and V. Kamenska, Symmetry and Information Content of Chemical Structures, *Bull. Math. Biol.*, **38**, 119–133 (1976).
20. E. Trucco, A Note on the Information Content of Graphs, *Bull Math. Biophys.*, **18**, 129–135 (1956); E. Trucco, On The Information Content of Graphs: Compound Symbols; Different States for Each Point, *Bull. Math. Biophys.*, **18**, 237–253 (1956).
21. D. Bonchev, O. Mekenyan, and N. Trinajstić, Isomer Discrimination by Topological Information Approach, *J. Comput. Chem.*, **2**, 127–148 (1981).

22. M. Gordon and J.W. Kennedy, *J. Chem. Soc. Faraday Trans.* II, **69**, 484–504 (1973).
23. J. R. Platt, Prediction of Isomeric Differences in Paraffin Properties, *J. Phys. Chem.*, **56**, 328–336 (1952).
24. S. Bertz, The First General Index of Molecular Complexity, *J. Am. Chem. Soc.*, **103**, 3599–3601 (1981).
25. D. Bonchev, and N. Trinajstić, Information Theory, Distance Matrix, and Molecular Branching, *J. Chem. Phys.*, **67**, 4517–4533 (1977).
26. D. Bonchev and N. Trinajstić, Chemical Information Theory. Structural Aspects, *Intern. J Quantum Chem. Symp.*, **16**, 463–480 (1982).
27. D. Bonchev, A. T. Balaban and 0. Mekenyan, Generalization of the Graph Center Concept, and Derived Topological Indexes. *J. Chem. Inf. Comput. Sci.*, **20**, 106–113 (1980).
28. D. Bonchev, A.T. Balaban and M. Randić, The Graph Center Concept for Polycyclic Graphs, *Intern. J. Quantum Chem.*, **19**, 61–82 (1981).
29. F. Harary, *Graph Theory*. Addison-Wesley, Reading, MA, 1969.
30. A.T. Balaban, Highly Discriminating Distance-Based Topological Index, *Chem. Phys. Lett.*, **89**, 399–404 (1982).
31. D. Bonchev, O. Mekenyan and A.T. Balaban, An Iterative Procedure for the Generalized Graph Center in Polycyclic Graphs, *J. Chem. Inf. Comput. Sci.*, **29**, 91–97 (1989).
32. S.C. Basak, A.B. Roy and J.J. Ghosh, Study of the Structure-Function Relationship of Pharmacological and Toxicological Agents Using Information Theory, in: *Proceedings of the Iind International Conference on Mathematical Modeling*, X, J.R. Avula *et al.* (Eds), University of Missouri, Rolla, Missouri, Vol. II, pp. 851–856.
33. V.R. Magnuson, D.K. Harris and S.C. Basak, Topological Indices Based on Neighborhood Symmetry: Chemical and Biological Applications, in: *Chemical Applications of Topology and Graph Theory*, R.B. King (Ed.), Elsevier, The Netherlands, pp. 178–191.
34. S.C. Basak, Information Theoretic Indices of Neighborhood Complexity and Their Applications, in: *Topological Indices and Related Descriptors in QSAR and QSPR*, J. Devillers and A.T. Balaban (Eds), Gordon and Breach, Reading, UK, pp. 563–593.
35. D. Bonchev, *Information-Theoretic Indices for Characterization of Chemical Structures*. Research Studies Press, Chichester, UK, 1983.
36. S.H. Bertz, Complexity of Molecules and Their Synthesis, in: *Mathematical Chemistry Series, Vol. VII Complexity in Chemistry*, D.

Bonchev and D.H. Rouvray (Eds), Taylor & Francis, London, UK, Chapter 3, herein.
37. S.H. Bertz and C.M. Zamfirescu, New Complexity Indices Based on Edge Covers, *Commun. Math. Chem.* (*MATCH*), **42**, 9–70 (2000).
38. S. Nikolić, N. Trinajstić, I.M. Tolić, G. Rücker, and Ch. Rücker, On Molecular Complexity Indices, in: *Mathematical Chemistry Series, Vol. VII Complexity in Chemistry*. D. Bonchev and D.H. Rouvray (Eds), Taylor & Francis, London, UK, Chapter 2, herein.
39. H, Wiener, Structural Determination of Paraffin Boiling Points, *J. Am. Chem. Soc.*, **69**, 17–20 (1947).
40. D. Bonchev, O. Mekenyan, and N. Trinajstić, Topological Characterization of Cyclic Structures, *Intern. J Quantum Chem.*, **17**, 845–893 (1980).
41. A.T. Balaban, D. Bonchev, X. Liu, and D.J. Klein, Molecular Cyclicity and Centricity of Polycyclic Graphs. I. Cyclicity Based on Resistance Distances or Reciprocal Distances, *Int. J. Quantum Chem.*, **50**, 1–20 (1994).
42. E. Ruch and I. Gutman, The Branching Extent of Graphs, *J. Combinatorics*, **4**, 285–295 (1979).
43. S.H. Bertz, Branching in Graphs and Molecules, *Discrete Appl. Math.*, **19**, 41–70 (1988).
44. D. Bonchev, Topological Order in Molecules. 1. Molecular Branching Revisited, *Theochem*, **336**, 137–156 (1995).
45. Randić, M. On Molecular Branching, *Acta Chim. Slovenica*, **44**, 57–77 (1997).
46. D. Bonchev, Shannon's Information in Assessing Structural Complexity, to be published.
47. D. Minoli, Combinatorial Graph Complexity, *Atti Acad. Naz. Lincei Rend. Cl. Sci. Fis. Mat. Natur.* (Ser. 8), **59**, 651–661 (1975).
48. D. Bonchev, O.N. Temkin, and D. Kamenski, On the Complexity of Linear Reaction Mechanisms, *React. Kinet. Catal. Lett.*, **15**, (1980) 119–124.
49. D. Bonchev, D. Kamensky, and O.N. Temkin, Complexity Index for the Linear Mechanisms of Chemical Reactions, *J. Math. Chem.*, **1**, 345–388 (1987).
50. I. Gutman, R.B. Mallion, and J.W. Essam, Counting the Spanning Trees of a Labeled Moleculae Graph, *Mol. Phys.*, **50**, 859–877 (1983).
51. P.E. John, R.B. Mallion, and I. Gutman, An Algorithm for Counting Spanning Trees in Labeled Molecular Graphs Homeomorphic to Cata-Condensed Systems, *J. Chem. Inf. Comput. Sci.*, **38**, 108–112 (1998).

52. S.H. Bertz, W.C. Herndon, Similarity of Graphs and Molecules, in: *Artificial Intelligence Applications in Chemistry*, American Chemical Society, Washington, DC, 1986, pp. 169–175.
53. S.H. Bertz and T.J. Sommer, Rigorous Mathematical Approaches to Strategic Bonds and Synthetic Analysis Based on Conceptually Simple New Complexity Indices, *Chem. Commun.*, 2409–2410 (1997).
54. S.H. Bertz and W.F. Wright, The Graph Theory Approach to Synthetic Analysis: Definition and Application of Molecular Complexity and Synthetic Complexity, *Graph Theory Notes of New York* (NY Acad. Sci.), **XXXV**, 32–48 (1998).
55. D. Bonchev, Kolmogorov's Information, Shannon's entropy, and Topological Complexity of Molecules, *Bulg. Chem. Commun.*, **28**, 567–582 (1995).
56. D. Bonchev, Novel Indices for the Topological Complexity of Molecules. *SAR QSAR Environ. Res.*, **7**, 23–44 (1997).
57. D. Bonchev, Overall Connectivity and Topological Complexity: A New Tool for QSPR/QSAR, in: *Topological Indices and Related Descriptors in QSAR and QSPR*, J. Devillers and A.T. Balaban (Eds), Gordon and Breach, The Netherlands 1999, pp. 361–401.
58. D. Bonchev, Overall Connectivities/Topological Complexities: A New Powerful Tool for QSPR/QSAR, *J. Chem. Comput. Sci.*, **40**, 934–941 (2000).
59. R.G.A. Bone and H.O. Villar, Exhaustive Enumeration of Molecular Substructures, *J. Comput. Chem.*, **18**, 86–107 (1997).
60. G. Rücker and Ch. Rücker, Walk Counts, Labyrinthicity and Complexity of Acyclic and Cyclic Graphs and Molecules, *J. Chem. Inf. Comput. Sci.*, **40**, 99–106 (2000).
61. D. Bonchev, Overall Connectivity—A Next Generation Molecular Connectivity, *J. Mol. Graphics Model.*, **5271**, 1–11 (2001).
62. D. Bonchev, The Overall Wiener Index—A New Tool for Characterization of Molecular Topology, *J. Chem. Inf. Comput. Sci.*, **41**, 582–592 (2001).
63. D. Bonchev, N.Trinajstić, Overall Molecular Descriptors. 3. Overall Zagreb Indices, *SAR QSAR Environ. Res.*, **12**, 213–235 (2001).
64. I. Gutman, B. Ruščić, N. Trinajstić and C.W. Wilcox, Jr., Graph Theory and Molecular Orbitals. 12. Acyclic Polyenes, *J. Chem. Phys.*, **62**, 3399–3405 (1975).
65. S. Nikolić, I.M. Tolić, N. Trinajstić and I. Baućić, On the Zagreb Indices As Complexity Indices, *Croat. Chem. Acta*, in press.
66. O. Mekenyan, D. Bonchev and N. Trinajstić, Chemical Graph Theory Modeling the Thermodynamic Properties of Molecules, *Intern. J.*

Quantum Chem. Symp., **18**, 369–380 (1980).

67. D. Bonchev, W.A. Seitz, C.F. Mountain and A.T. Balaban, Modeling the Anticarcinogenic Action of Some Retinoid Compounds by Making Use of the OASIS Method. III. Inhibition of the Induction of Ornithine Decarboxylase by Arotinoids, *J. Med. Chem.*, **37**, 2300–2307 (1994).

Chapter 5

THERMODYNAMIC COMPLEXITY

W.A. Seitz
Program for the Theory of Complex Natural Systems,
Texas A&M University at Galveston, Galveston, TX 77553, USA

5.1 Introduction

Complexity is the latest in a series of science concepts to achieve widespread recognition both in the scientific and popular scientific literature. Predecessors include catastrophe theory [1], fractal geometry [2] in nature, cellular automata [3], non-linear dynamics [4] and chaos theory [5]. Each of these concepts has claimed intellectual breakthroughs, led to establishment of new areas of research (evidenced by journals with corresponding titles), and attracted a wide range of people to study the new horizons each viewpoint exposed. The latest such concept, *complexity*, has even captured the attention of non-scientists such as military strategists, economists and stock market analysts.

In science, the complexity notion has proved helpful in describing phenomena such as self-organization in physicochemical systems and biological evolution. One of the early more readable descriptions of complexity in chemistry and physics is the book *Exploring Complexity* by

Nicolis and Prigogine published over a decade ago [6]. In the ensuing time, while the concept continues to be applied in ever-wider circles, there has been limited success in providing a quantitative measure to it. Complexity is still chiefly a concept or point of view and not a well-defined property of matter or processes. In the words of Tor Nørretranders, "Today we have no quantitative notion of complexity—i.e., a notion that allows us to measure complexity. So it is not yet an area that counts for anything among physicists" [7].

Here we discuss complexity in steady-state chemical systems and attempt to begin quantification of the term for simple cases. In more "complicated" systems, these notions can be expanded into a sequence of "steps" resulting from modifications of the steady-state condition implemented via local interactions with the environment. This latter sequence resembles Bennett's concept of logical depth [8], and Lloyd and Pagels' [9] notion of thermodynamic depth.

In this chapter I first describe briefly some previous applications of complexity concepts. Next I propose a chemical thermodynamic state function for complexity. To illustrate the function, I consider it in the context of phase equilibrium where complexity is computed (defined) for liquids. The discussion is carried out in the context of Maxwell's Demon and the concept of microscopic (or internal) work. Finally, complexity is computed quantitatively for near-equilibrium steady state ideal gases subjected to a small thermal gradient.

5.2 The Emergence of Complexity

Waldrop recently wrote a popular book entitled *Complexity* [10] that described the evolution of a group of philosophers, economists, and scientists. These individuals included Nobel Prize winners, economists, geologists, physicists, computer scientists interested in evolutionary biology and evolutionary biologists interested in computer simulations. They came together (much as was the case in the earlier development of chaos theory) outside of, and independent of their "home" institutions or disciplines. A major center in the United States today for this area of research (and much of chaos theory as well) is the Santa Fe Institute in New Mexico.

Several examples of complexity were explored in Wadrop's book, and since then many more have been discussed. Examples from the fields of biology, chemistry, geosciences, and economics have been cited in the literature. A recent issue of Science [11] provided numerous examples of complexity in these fields. In biology, complexity concepts are invoked in discussions of evolution, emergence of order from disorder, and even consciousness. Kaufmann in his book *Origins of Order* [12] introduces a wide variety of computer simulations from which ordered structures emerge as the result of the interconnectivity of system parts as opposed to details of the interconnections themselves. The notion of evolution of structure is also discussed widely by scientists interested in neural networks and their ability to "learn" through interactions with an environment (or training set). In chemistry, the simplest "complex" structure usually identified is the Bénard circulation pattern that emerges when fluids are subjected to sufficiently high thermal gradients (but less than that which promotes turbulence). The famous ZBT chemical oscillator is another chemically complex system widely cited. Even military strategy has not been exempt from the impact of complexity theory. The latest naval strategy is based on "speed of command" and "sensor input" allowing field commanders greater flexibility of actions with the end result being an evolving (but now interconnected via data sharing) battlefield which rewards fast response (evolution) with victory [13].

But despite this widespread use of the term complexity, without quantification, it will fail to win the acceptance among scientists except as another descriptive word (as Nørretranders pointed out). Several approaches have been undertaken in the past. One approach has been the use of information-theoretic approaches to compute the information content in molecular structures. Second, entropy measures have been applied, with complexity related in some fashion to the negative of the entropy. Third, Prigogine and others study "dissipative structures", their evolution and their non-equilibrium thermodynamic properties. This latter approach has had some success in focusing on energy fluxes necessary to support dissipative structures. All of these approaches either describe the complexity of an object (e.g., a polymer or complex organism) or of a process—two quite different viewpoints. However, given both the enormous range of subjects in which the term (and the general concept) complexity is utilized, it is not surprising that no single definition has emerged.

5.3 Object Complexity

Long before the term complexity became popular as a description for chemical systems, a variety of theories were developed to quantify the topology of objects. These views all take as a starting point an object (or set of objects) that are then analyzed using a variety of techniques.

In chemistry, the molecular formula provides one type of "measure" for the complexity of an object. The rules for nomenclature developed by IUPAC describe a molecule in the sense that the name can be used to determine the structure and vice versa. As compounds of interest to chemists have become increasingly large, the limitations of nomenclature for defining complexity have become clear. Nevertheless, nomenclature remains an important tool in chemistry.

Kolmogorov [14] was the first to provide a generally accepted measure of the complexity of an object. In the 1960s he proposed that an object's complexity increases as the length of the shortest description for the object. His approach, while similar to rules of nomenclature, greatly generalized the concept to include practically all types of "descriptors." Kolmogorov thus provided a consistent and appealing definition, which continues in use today. Critics of Kolmogorov complexity point out (correctly) that the most complex objects under this definition are those with the most randomness. Hence, white noise is more complex than a symphony. Because of this feature, chemists and others continued to seek other definitions for complexity capable of discerning the features of greatest interest to them. Nevertheless, Kolmogorov complexity—as a concept—is both consistent and useful.

A specific example for Kolmogorov entropy is as follows. One "description" for molecular structure arises from the so-called adjacency matrix in which the $(i,j)^{th}$ element of the matrix is 1 if atoms i and j are connected by a bond and 0 otherwise. Given this matrix one can define unique molecular descriptors at the level of detail contained in the matrix itself. Given the descriptors (in general a sequence of 1's and 0's) Kolmogorov complexities based on the adjacency matrix can be computed simply using standard Boolean techniques [15]. The point to be made again is that this approach, while general and useful, does not necessarily expose all of the features of complexity of interest to chemists.

5.4 Process Complexity

The use of the term complexity to describe processes that form structures rather than the structures themselves is more recent than the object complexity discussed earlier. Nevertheless, this form of complexity has led to developments and applications of the concept that have provided insight into many diverse fields.

Complexity can arise out of randomness whenever a physical system is externally forced from its equilibrium state. The two best known examples of this in chemistry are the formation of convection cells when confined fluids are subject to a temperature gradient, and oscillatory chemical reactions where the forcing is via continuous mass flow. In both cases a system initially at equilibrium is perturbed, with the perturbation being continuous and leading to a new non-equilibrium stationary state exhibiting structure or complexity. The maintenance of the complex structure requires a continuing flux of energy or matter.

We recently proposed a generalization of Le Chatelier's principle: "It is a feature of non-equilibrium systems that spontaneous processes induced by a deviation from the initial state proceed in such a manner as to select states most effective in preserving system properties" [16]. In a less general form, Odum proposed a similar theory [17] for long-term evolutionary "strategy" of the biosphere—"namely increased control or homeostasis with the physical environment in the sense of achieving maximum protection from its perturbations."

Another example of biological evolution under external stress, i.e. subject to perturbation (mentioned briefly earlier) is Kauffman's treatment of complex biological "landscapes" [18]. Kauffman considers the formation of proteins from their amino acid components. A specific protein is being sought that carries out some biological function optimally—e.g., an antibody effective against a specific antigen. Here the driving forces are two-fold—first the antigen itself perturbs the organism threatening it with death, and second, energy is available from respiration or photosynthesis permitting complex (non-equilibrium) molecule formation. Organisms that effectively resist antigens survive. The search for an optimal antibody takes place on a multi-dimensional "fitness landscape" with the altitude of each point being the effectiveness of the particular protein to resist the antigen, and where the geography is all proteins. Thus, proteins that are similar in structure are nearby geographically, but can have wildly varying altitudes. In general, the landscapes are very rough. To search for an optimum, energy is provided to build proteins (i.e., to visit different geographic regions) and then to replicate those with high fitness. This so-called genetic algorithm for optimization is now generally applied in mathematics and computer science [19].

Prigogine's book *Exploring Complexity* is almost entirely devoted to process complexity and focuses on the non-equilibrium thermodynamics associated with the energy or mass fluxes required to maintain the complex process. But the "thing" being maintained can also be viewed independently of the flux and its complexity analyzed via the methods discussed earlier in the section on object complexity. What is lacking is a quantitative connection between the process that generates the object and the object itself.

In what follows, I will propose a chemical thermodynamic complexity function and compute the complexity of a steady-state system (i.e. object), namely an ideal gas subjected to a small thermal gradient. By doing this, we see how the complexity computed directly from standard statistical mechanics can be applied to a steady-state object (a thermal gradient) and then connected to the process which created it (heat flux).

Before beginning on non-equilibrium systems (simple though they may be), I will digress a bit into the more difficult area of dynamic equilibrium between liquids, solids and gases. Here, while the overall system may be at equilibrium, there is a constant flux of matter between the phases—that flux averaging to zero. The discussion is intended to introduce the thermodynamic complexity function in terms familiar to chemists, but not generally treated in thermodynamic texts.

5.5 Chemical Context

Thermodynamics is the study of systems near to equilibrium. It forms a central part of every chemist's heritage. The well-known Carnot engine is one of the more elegant "thought experiments" based on our quantitative understanding of the behavior of ideal gases both in macroscopic and in microscopic statistical terms. The thermodynamic function, entropy, which emerges both statistically and macroscopically, is one of the subtlest quantitative properties of matter. It is no wonder that many attempts to define complexity have close ties to the opposite of entropy—with some authors even proposing that the complexity can simply be regarded as the negative of the entropy [20]. Shannon's information theory [21] also utilizes the statistical functional form for entropy to describe the information content of a system. Some authors propose that complexity be regarded as that value. But as discussed above, these approaches lead to the conclusion that the most complex systems are those that possess the most disorder.

When a system is near to equilibrium, its properties do not change significantly with time. It is then possible to view the local behavior of the particles (molecules) making up the system as being the same throughout the whole. Thus, near equilibrium particles in the system experience a "context." In ideal gases, this context is simply a free particle subject to occasional collisions, real gases are similar, but with some interaction corrections. In liquids the context is much richer and measured usually in terms of radial distribution functions that are in general quite complicated. In solids the context is again quite simple in that the particles are no longer free to move about. A rough graph of chemical context as a function of state is given in Figure 5.1 (where the ordinate is some measure of the length of the description for the context over some brief time period).

The notion that somehow complexity is found between order and disorder has a long history. In fact, the graph above is identical to that of Huberman and Hogg's complexity graph [22] if one replaces the terms solid and gas with order and disorder respectively. Thus if we were to ask which phase of matter is the most complex (for a pure substance), the answer suggested by the chart above is the liquid state.

The notion of chemical context above may apply in discussions of the requirement for Maxwell's Demon to achieve violations of the second law of thermodynamics. Consider the demon in its usual form, namely a being capable of opening and closing a portal between two regions of a container of gas at temperature T without the expenditure of energy. If s/he is capable of knowing whether a given particle approaching the portal is more or less energetic than the average, s/he might accomplish separation of particles according to their energy. It is now recognized that the demon fails to separate fast from slow moving gas particles due not to his/her inability to know all of their positions and momenta (i.e. energy and position), but rather to her/his inability to keep track of this information through time—the latter requiring either infinite storage capacity for the demon or require work to discard information.

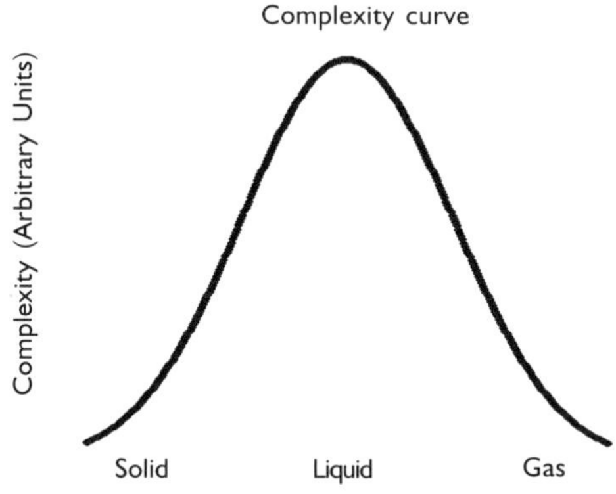

Figure 5.1

Now, suppose the demon were employed to liquefy a gas near its condensation point, s/he would move particles from a relatively simple environment and place them in an increasingly complex one—a task requiring increased effort the greater the quantity of liquid. On the other hand, to solidify a liquid near its freezing point, s/he must keep less and less information regarding individual particles as time goes on since the amount of the complex phase is decreasing. Thus in a quantitative sense, the difference between energy of fusion and energy of vaporization is related to the complexity of the liquid.

To further explore this idea in thermodynamic terms consider the usual form for the first law

$$dE = dq + dw \tag{5.1}$$

where dq is the heat added to the system, dw is the work done on the system and dE is the change in energy of the system. Now textbooks set

$$dw(\text{system}) = -dw(\text{surroundings}) \tag{5.2}$$

i.e. they assume that if no work is done on or by the surroundings, then no work is done on or by the system. While this provides a consistent view, it is not necessary to assume this. In fact, if one divides equation (5.1) by the temperature T one obtains for reversible processes

$$\begin{aligned} dE/T = dq_{rev}/T + dw_{rev}/T &= dS + dw_{rev}/T \\ &= dS + d\mathit{W} \end{aligned} \tag{5.3}$$

Where dS is the entropy change and $d\mathit{W}$ is a thermodynamic function (not usually studied in texts). If we do not assume equation (2) but rather allow for the concept of internal work being done on a system via infusion of energy (heat, mass or other) we can define a state function for complexity

$$dC = dw_{rev} = dE - T\,dS \tag{5.4}$$

For normal heating processes, the change in energy upon addition of heat is accompanied by a change in temperature, with the heat capacity containing details of the properties of the system undergoing the change. At phase transitions (specifically fusion and vaporization) the heat capacity of all substances is infinite and there is no temperature change with the introduction of energy to the system. If one assumes that the energy input during a phase transition is available to a demon to do work as opposed to entropy production/reduction—namely breaking bonds or reducing intermolecular interactions, one can propose the following

$$\Delta H_{trans} = \Delta C_{trans} \tag{5.5}$$

Then, if one takes the complexities of gas and solid as zero (Figure 5.1), we find that

$$[\Delta H_{vap} - \Delta H_{fus}] = C \text{ (liquid)} \tag{5.6}$$

Thus, the difference between heats of fusion and vaporization provides a quantitative measure of liquid complexity. Table 5.1 gives some values for the heats of fusion and vaporization for a set of molecules and computes the "liquid complexity" as their difference.

As a qualitative test of this idea, consider the following heuristic argument that compares the complexity of liquids. Elemental liquids are expected to be the least complex since the only information required specifying the liquid state (instantaneously) is the position of the molecular centers. Diatomic liquids are more complex due to the requirement of specifying both position and orientation of the particles (richer chemical context).

Table 5.1: Liquid Complexities C

Compound	ΔH_{fus}	ΔH_{vap}	**C (liquid)**
Hydrogen	0.12	0.904	0.784
Nitrogen	0.72	5.577	4.857
Oxygen	0.444	6.82	6.376
methane	0.941	8.18	7.239
ethane	2.86	14.72	11.86
n-octane	20.74	34.98	14.24
propane	3.52	18.78	15.26
n-hexane	13.03	28.85	15.82
Sulfur dioxide	7.4	24.92	17.52
Ammonia	5.65	23.33	17.68
n-butane	4.66	22.39	17.73
Chloroform	9.2	29.4	20.2
Carbon Tet.	2.5	30	27.5
Methanol	3.17	35.27	32.1
Ethanol	5.02	38.58	33.56
Water	6.01	40.66	34.65
Nitro-benzene	11.59	55.01	43.42

Triatomic molecules might be even more complex in their liquid state due to the greater number of configurations accessible, etc. … Thus to zeroth order, we might say that the complexity of a liquid varies with the number of atom types making up the molecule. This argument is the basis for Figure 5.2 below, where the points plotted are averages of the data in Table 5.1, vs. atom types in the molecule. Certainly, the heuristic argument is oversimplified (e.g., note the position of water), but the surprising linear relationship of the averages in Table 5.1 to the heuristic argument suggests that there may be further correlations possible given a more detailed theory of chemical context. In other words, the chemical context involves more than just the orientations of individual particles; there are also interactions that need be considered in a more detailed analysis.

In the above, the energy input during a phase change was regarded as being available to Maxwell's Demon for microscopic work. This view will be expanded further below where we consider non-equilibrium, but steady state systems (see note 1).

5.6 Quantifying Complexity in Thermal Gradients

We come finally to the promised quantification of complexity for a simple (simplest?) non-equilibrium system—namely a gas contained between two heat baths of differing temperature (Figure 5.3). The thermal gradient ΔT is not sufficient to produce macroscopic fluid flow, and only

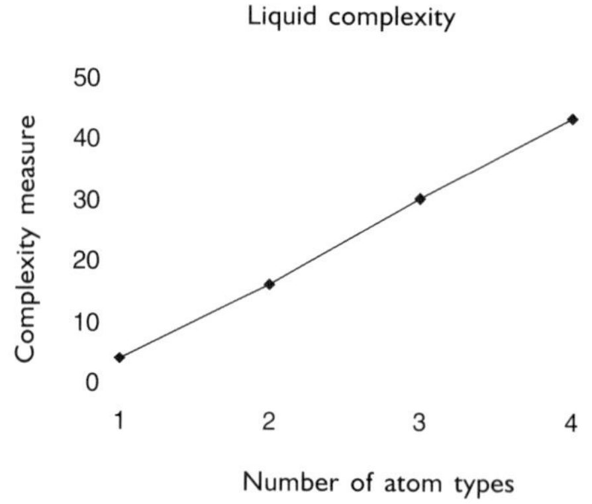

Figure 5.2: Chart of $C = [\Delta H_{vap} - \Delta H_{fus}]$ vs. atom type in Table 5.1.

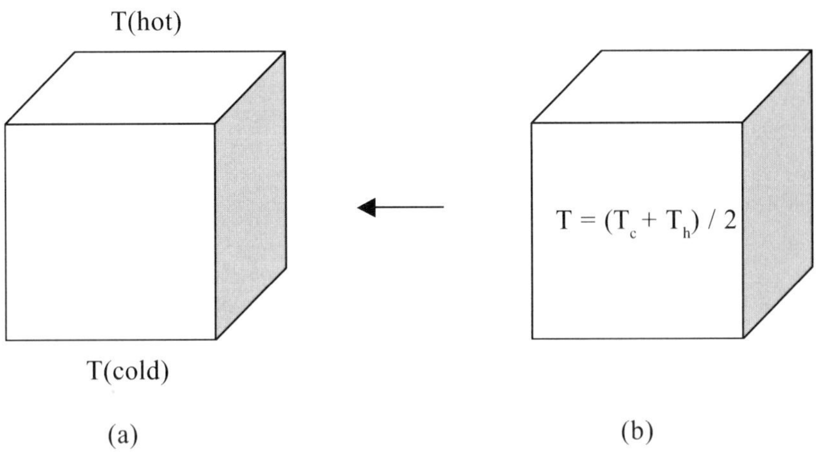

Figure 5.3: Ideal gas subjected to a temperature gradient.

produces a uniform thermal gradient in the gas. This is a steady-state system maintained by continuous passage of heat from the hot to cold reservoir—this heat supplying the necessary energy to maintain the gradient. Figure 3(b) simply shows the equilibrium state for the same gas at a temperature midway between the hot and cold reservoir.

It is easy to compute ΔE and ΔS for the systems in Figure 5.3. ΔE is strictly zero since for ideal gases the internal energy depends only on the temperature and the "average" temperature of the two systems is identical. However, ΔS is not zero!

For ν moles of an ideal (monatomic) gas at temperature T and volume V one has [23]

$$S(T, V, \nu) = \nu\{C_V \ R \ln(T) + R \ln(V) - R \ln(\nu)\}$$

and for an ideal monatomic gas with P =1 and C_V = 3/2R this becomes

$$S(T, V) = \nu\{5/2 \ R \ln(T) + R \ln(R)\} \tag{5.7}$$

Now to compute the entropy of the thermal gradient system (3a) (for one mole of gas) one simply integrates equation (5.7).

$$S = \int_{\nu=0, T=Tcold}^{\nu=1, T=Thot} \left(\tfrac{5}{2} R \ln(T) + R \ln(R)\right) d\nu$$

For an infinitesimal change in T, dT, there is a corresponding number of moles Δdv where $\Delta = (T_{hot—}T_{cold})$ so that the integral becomes.

$$S = \frac{5R}{2\Delta}\{T_{hot}\ln T_{hot} - T_{cold}\ln T_{cold} - \Delta\} + R\ln R \tag{5.8}$$

Subtracting (5.7) from (5.8) yields

$$\Delta S = 5/2R\{\ln(T_{cold} + \Delta/2) - 1/\Delta[(T_{cold} + \Delta)\ln(T_{cold} + \Delta) - T_{cold}\ \ln T_{cold} - \Delta]\} \tag{5.9}$$

Two plots of equation (5.9) are given below. Figure 5.4 shows the results if T_{cold} is low. While strictly speaking the slope of ΔS at the origin is zero, very near the origin, ΔS behaves linearly. This behavior can be further analyzed in terms of linear response theory, but this is left for further work.

Inspection of Figure 5.5 indicates an inflection point. Upon setting $d\Delta S^2/d\Delta^2 = 0$ one obtains the equation

$$\frac{\Delta^2}{4(T_{cold} + \Delta/2)^2} + \frac{\Delta}{T_{cold} + \Delta} + \frac{2T_{cold}}{\Delta}[\ln(T_{cold} + \Delta) - \ln T_{cold}] - 2 = 0$$

which may be solved numerically. The physical significance, if any, of this inflection point is not known, but we do note that real fluids subjected to thermal stress make a transition from a uniform temperature gradient to a Bénard flow if Δ is sufficiently large.

Now equation (5.9) gives the entropy difference between the thermal gradient steady state gas and the equilibrium gas of averaged temperature. But since ΔE = 0, we have from equation (5.3) that

$$\begin{aligned} 0 &= \Delta E = T\Delta S + \Delta C \\ \Delta C_{\text{thermal gradient}} &= -T\Delta S \end{aligned} \tag{5.10}$$

which achieves the quantification of complexity for a steady state thermal gradient. In the following section we provide some conclusions based on this description and discuss possible implications in the evolution of non-equilibrium systems.

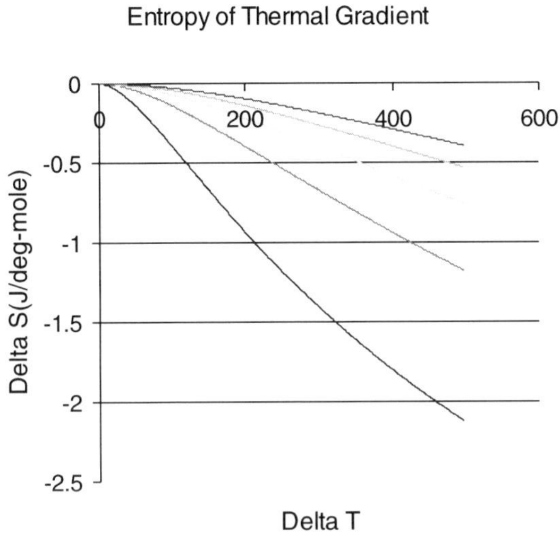

Figure 5.5: T_{cold} is 100 for the lowest curve and increases to 200, 300, 400 and 500. Note the inflection point.

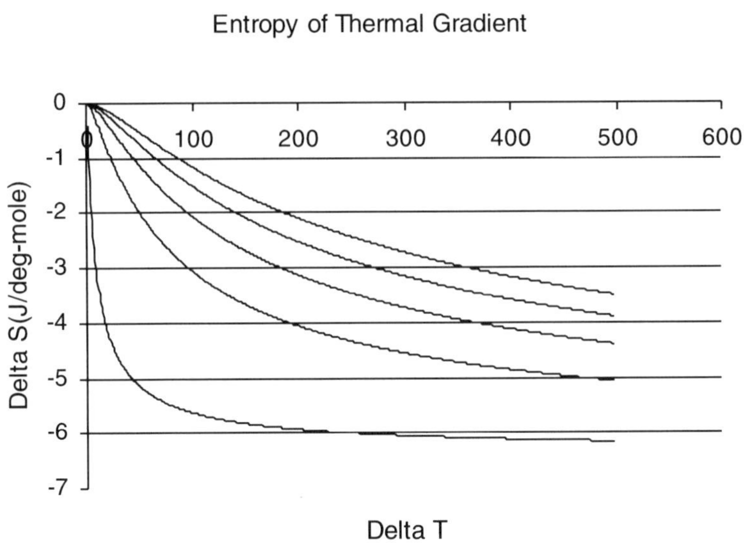

Figure 5.4: T_{cold} is 1°K for the lowest curve. They increase to 11, 21, 31 and 41.

5.7 Complexity, Entropy and Evolution

The arguments leading to the conclusion in equation (5.10) were based on the assumption that dw_{rev} can be non-zero despite the fact that no work is done on or by the surroundings. This conversion of entropy to work underlies some other approaches to understanding the evolution of structure in systems that "export" entropy. This available entropy can then be utilized by the system (Maxwell's Demon) to establish structure (such as the thermal gradient). Now, unless there is some violation of the second law (which is certainly not being proposed), we have the inequality (for irreversible systems):

$$\Delta S_{universe} = \Delta S_{system} + \Delta S_{surroundings} > 0$$

so that the flow of heat from hot to cold reservoir required to generate the temperature gradient must generate an increase in entropy at least as great as the entropy decrease associated with the gradient itself.

However, the system (even this simple steady state thermal gradient) can only be maintained by a continuous supply of energy in time. This topic has been extensively and productively explored by Prigogine and others who study energy fluxes as being associated with complexity. The view taken here, however, is a static one—namely one that computes directly an entropy difference between a steady state structure and an equilibrium system.

To see how this view might be utilized in practice, we must consider the details of the system being driven by the external energy flow and observe or compute the relaxation time between the steady state and equilibrium or vice-versa. This then provides the time required for the given energy flux to produce the structure and is related to the complexity. Once produced, however, the structure does not undergo further organization despite the continued export of entropy. In other words, the relaxation time from the steady state structure to equilibrium (or the time required to produce the steady-state system beginning from equilibrium conditions) provides a bound for the complexity of the structure if one can compute the entropy flux to the environment. Since the latter is sometimes more easily computed than the computation of the entropy difference itself, the concept can prove useful. For steady states where $\Delta E = 0$, we have

$$\Delta C <\!\!-\!\!-T\Delta S = T * (\text{entropy flux to the surroundings})(\text{relaxation time}) \tag{5.11}$$

where the inequality is that from the second law. Thus a measure of the entropy flux to the surroundings coupled with an estimate of the relaxation time of the "object" provides a measure of the maximum complexity of the system.

We end this article with some speculations regarding the extension of these ideas to the limits of evolution in large systems such as organisms, ecosystems and planets. The earth as a system exports entropy through the conversion of high energy photons to low energy photons. This entropy export is the result of the following. Assume the earth emits approximately the same amount of energy as is absorbed (i.e. no net heating is observed over time due to absorption of solar radiation). Then we have that ΔE is approximately zero for the earth. However, the absorbed photons (UV and visible) are of much higher frequency than the emitted photons (IR). Since as much energy must be radiated at low frequencies as is absorbed at high frequencies more photons must be emitted than absorbed. The result is a significant export of entropy from the earth system. Thus, the earth system is one that is capable of complexity creation without violation of the second law. However, equation (5.11) provides a limit to the complexity that can be created.

However, we note that a given entropy flux (e.g., earth radiation) is capable of producing more and more complex objects if the corresponding relaxation times of those objects also increase. It is interesting that the characteristics of the "mature stage" of ecosystems proposed by Odum and others is that organisms are of large size with long and complex life cycles. These ideas tie to Pagels and Lloyd's notion of thermodynamic depth that is "the idea of defining complexity as the amount of information that is discarded *during the process* that brings a physical object into being." According to Lloyd and Pagels, "Complexity must be a function of the process —the assembly routine—that brought the object into existence." In particular, for the simple system considered here the chief characterization of the "process" is the relaxation time.

In conclusion, I have attempted to provide a quantitative approach for measurement of complexity for simple systems near to equilibrium. Here the process simply involved particle redistribution via Maxwell's Demon with the work provided by the decreased entropy. Further work is necessary to attack "more complex" objects further from equilibrium—perhaps following a series of steps (the assembly routine) and further quantifying thermodynamic depth. We have found a connection between object and process complexity, but much remains to be done. In particular, the kinetics of "evolution" of systems subject to entropy flux remains for future work.

Acknowledgements

The author acknowledges the support of the Welch Foundation.

Note

[1] It is interesting to note that phase equilibrium is regarded as a special type of equilibrium, namely a dynamic equilibrium where individual molecules move between phases, but on average there is no net change in the phase fractions unless energy is provided or removed. In some respects this is similar to steady state systems.

5.8 References

1. J. Moser, *Stable and Random Motions in Dynamical Systems*, Princeton University Press, Princeton, NJ, 1973.
2. B. Mandelbrot, *Fractals: Form, Chance, Dimension*, Freeman, San Francisco, 1977.
3. S. Wolfram (Ed.), *Theory and Applications of Cellular Automata*, World Scientific, Singapore, 1987.
4. P. Collet and J.P. Eckmann, *Iterated Maps on the Interval as Dynamical Systems*, Birkhauser, Basel, 1980.
5. H. Shuster, *Deterministic Chaos*, Physik-Verlag, Weinheim, 1984.
6. G. Nicolis and I. Prigogine, *Exploring Complexity*, Freeman, New York, 1989.
7. T. Nørretranders, *The User Illusion*, Penguin Putnam, New York, 1998.
8. C.H. Bennett, Logical Depth and Physical Complexity, in: *The Universal Turing Machine. A Half Century Survey*, Rolf Herken, Oxford University Press, Oxford, 1990.
9. S. Lloyd and H. Pagels, Annals of Physics, **188**, 186–213 (1988).
10. M.M. Waldrop, *Complexity: The Emerging Science at the Edge of Order and Chaos*, Simon and Schuster, New York, 1992.
11. *Science*, **284**, (1999).
12. S.A. Kaufmann, *The Origins of Order: Self-Organization and Selection in Evolution*. Oxford University Press, New York, 1995.
13. A.K. Cebrowski, Network Centric Warfare: A Revolution in Military Affairs, presented at the Current Strategy Forum, US Naval War College, 1997.
14. A.N. Kolmogorov, *Problems in Information Transmission*, **1**(1) (1965); R.J. Solomonoff, *Information and Control*, **7**(22) (1964); G.J. Chaitin, *J. Assoc. Comp. Mech.*, **16**, 145, (1969).
15. D. Bonchev—to be published.

16. D. Bonchev and W.A. Seitz, The Concept of Complexity in Chemistry, in: *Concepts in Chemistry: A Contemporary Challenge*, D.H. Rouvray (Ed.), John Wiley and Sons, New York, 1997.
17. E.P. Odum, *Science*, **164**, 262 (1969).
18. S.A. Kauffman, *At Home in the Universe: The Search for the Laws of Self-Organization*, Oxford University Press, New York, 1993.
19. S. Forest, *Science*, **261**, 872 (1993).
20. S. Forest, *Science*, **164**, 262 (1993).
21. E. Schrödinger, What is Life from the Standpoint of Physics, Cambridge University Press, New York, 1992.
22. C. Shannon and W. Weaver, *Mathematical Theory of Communications*, University of Illinois Press, Urbana, 1949.
23. B.A. Huberman and T. Hogg, *Physica*, **22D**, 376 (1986).
24. F. Reif, *Fundamentals of Statistical and Thermal Physics*, McGraw-Hill, New York, 1965.

INDEX